Silke Bollmohr

An Exposure and Effect Assessment of Insecticides in an Estuary

Silke Bollmohr

An Exposure and Effect Assessment of Insecticides in an Estuary

Spatial and temporal variability of particle bound insecticides in a temporarily open estuary, South Africa

Südwestdeutscher Verlag für Hochschulschriften

Impressum/Imprint (nur für Deutschland/ only for Germany)
Bibliografische Information der Deutschen Nationalbibliothek: Die Deutsche Nationalbibliothek verzeichnet diese Publikation in der Deutschen Nationalbibliografie; detaillierte bibliografische Daten sind im Internet über http://dnb.d-nb.de abrufbar.
Alle in diesem Buch genannten Marken und Produktnamen unterliegen warenzeichen-, marken- oder patentrechtlichem Schutz bzw. sind Warenzeichen oder eingetragene Warenzeichen der jeweiligen Inhaber. Die Wiedergabe von Marken, Produktnamen, Gebrauchsnamen, Handelsnamen, Warenbezeichnungen u.s.w. in diesem Werk berechtigt auch ohne besondere Kennzeichnung nicht zu der Annahme, dass solche Namen im Sinne der Warenzeichen- und Markenschutzgesetzgebung als frei zu betrachten wären und daher von jedermann benutzt werden dürften.

Verlag: Südwestdeutscher Verlag für Hochschulschriften Aktiengesellschaft & Co. KG
Dudweiler Landstr. 99, 66123 Saarbrücken, Deutschland
Telefon +49 681 37 20 271-1, Telefax +49 681 37 20 271-0, Email: info@svh-verlag.de
Zugl.: Landau, Universität Koblenz-Landau, Dissertation, 2008

Herstellung in Deutschland:
Schaltungsdienst Lange o.H.G., Berlin
Books on Demand GmbH, Norderstedt
Reha GmbH, Saarbrücken
Amazon Distribution GmbH, Leipzig
ISBN: 978-3-8381-0846-9

Imprint (only for USA, GB)
Bibliographic information published by the Deutsche Nationalbibliothek: The Deutsche Nationalbibliothek lists this publication in the Deutsche Nationalbibliografie; detailed bibliographic data are available in the Internet at http://dnb.d-nb.de.
Any brand names and product names mentioned in this book are subject to trademark, brand or patent protection and are trademarks or registered trademarks of their respective holders. The use of brand names, product names, common names, trade names, product descriptions etc. even without a particular marking in this works is in no way to be construed to mean that such names may be regarded as unrestricted in respect of trademark and brand protection legislation and could thus be used by anyone.

Publisher:
Südwestdeutscher Verlag für Hochschulschriften Aktiengesellschaft & Co. KG
Dudweiler Landstr. 99, 66123 Saarbrücken, Germany
Phone +49 681 37 20 271-1, Fax +49 681 37 20 271-0, Email: info@svh-verlag.de

Copyright © 2009 by the author and Südwestdeutscher Verlag für Hochschulschriften Aktiengesellschaft & Co. KG and licensors
All rights reserved. Saarbrücken 2009

Printed in the U.S.A.
Printed in the U.K. by (see last page)
ISBN: 978-3-8381-0846-9

Nothing has changed!!

"…….The paradox between attempting to analyze "too much" information and still not having enough – although frustrating – should not be discoursing, for this will lead to eventual acknowledgement by our administrators that complex problems do not have simple solutions. This is progress. Biology without pollution is intricate, exacting and dynamic, while biology compounded by a single source of pollution may at times be overwhelming. Thus, biology with multiple-variable pollutants demands extraordinary insight as well as foresight into placing the problems into perspective….."

(Salo 1974)

Acknowledgement

This thesis is the result of research undertaken during the German-South African collaboration between University Koblenz-Landau and University of Cape Town. I gratefully acknowledge funding received from the Volkswagen Stiftung Hannover, the Graduiertenförderung of Lower Saxony, Germany and the Ministry for Science of Rhineland-Palatinate, Germany.

This section of my thesis should probably be the most read, as there are so many people who have helped me on the road to my dissertation who require special mention. First of all to my supervisor, Ralf Schulz, despite appearing to be overloaded with work you made yourself available with little fuss and helped me in any sense to finish the thesis. For this support and belief as a friend and supervisor in my potential I am truly grateful. To my co-supervisor Jenny Day, you supported me with finances, space and equipment and to give me inspiration and humor whenever I needed it. For your long belief in my potential I am enormously thankful.

I would also like to thank Sue K.C. Peall and Christina Hahn for sample preparation and pesticide analysis and the management of AECI, Somerset-West, for permission to enter their properties.

Special thanks to all the students, especially Maren Gelhar and Nadja Sahm, for assisting me in the very time and strength consuming field sampling. I would also like to thank Prof. Dr. Otto Larink, who always showed interest in my work

I really enjoyed the companionship of my UCT colleagues and friends who made me feeling home in Cape Town: Charlene Coulsen (various coffees and cigarettes), Maria and Kristin (long inspirational talks), Bruce (calm intensive discussions), Nadja Sahm (parties and living together), Andrea Plos (helping in any IT or infrastructure related problems), James Dabrowski and many many more students.

Unfortunately, I left some people behind in Germany which was definitely note easy in the beginning. However, all of you were continuously on my mind and I would like to thank especially Sylvia and Istrid (my sister) for their unconditional love, friendship and understanding.

During the last part of the thesis, I met my soul-mate, Markus. Thank you so much for your continuous understanding in difficult times, love and deepful discussions with many bottles of wine. I love you.

Also special thanks to my daughter, Nasrin. You gave me a lot of energy, support (e.g. dishwashing) and joy to continue with the thesis.

I greatly appreciate the patience and support from my parents, and I would like to express my enormous grateful thanks to my mom for her unconditional love and support in finances, time and inspiration. Without you, there would not have been this thesis. ***That's why I would like to dedicate the thesis to my mom Sigrid Bollmohr.***

TABLE OF CONTENTS

List of Tables		i
List of Figures		iii
Summary		vi
Zusammenfassung		x
Chapter 1	General Introduction	1
Chapter 2	Study area	15
Chapter 3	Temporal variability in particle-associated pesticide exposure in a temporarily open estuary, Western Cape, South Africa.	21
Chapter 4	Spatial and temporal variability in particle associated pesticide exposure and their effect on benthic community structure along a temporarily open estuary	39
Chapter 5	Interactive effect of salinity decrease, salinity adaptation and chlorpyrifos exposure on an estuarine harpacticoid copepod, *Mesochra parva*, in South Africa	60
Chapter 6	Comparison of environmental variables, anthropogenic stressors and benthic community structure in two temporarily open estuaries differing in size and type of catchment	79
Chapter 7	General Discussion	100
Chapter 8	References	109

List of Tables

Table 2.1 Site selection with its coordinates and the use in various chapters. 18

Table 2.2 Site selection within the Rooiels River estuary with its coordinates and the use in various chapters. 19

Table 3.1 Mean (± standard error) seasonal fluctuation in physicochemical characteristics measured in the Lourens River within the period 2001-2003. 27

Table 3.2 Mean (± standard error) particle-associated pesticide concentrations (µg/kg) in the Lourens River estuary per season and 90th percentile of all values within the overall sampling period 2001-2003. 29

Table 3.3 Chemical properties (ARS, 1995; ATSDR, 2000; Pesticide Manual) of the pesticides detected in the Lourens River estuary 30

Table 3.4. Estimation of mean (50% confidence) hazardous concentration which affects 5% of freshwater ($HC5_{freshwater}$) and marine organisms ($HC5_{marine}$) derived from species sensitivity distribution for arthropods and fish separately exposed to pesticides in single-species acute tests (US EPA database AQUIRE) by using the programme BurrliOz (* not sufficient data to perform analysis). 32

Table 3.5. Water and sediment quality criteria set by various international guidelines in comparison to HC5 concentration derived in this study for freshwater and marine organisms. 34

Table 3.6. Calculated chronic and acute toxicity exposure ratio (TER) for the pesticides detected in the Lourens River estuary using the calculated concentrations in the water and the calculated predicted no-effect-concentrations. 35

Table 4.1 Physicochemical parameters measured within the upper and middle reaches of the Lourens River estuary (Only the variables which are spatially significantly different ($p<0.05$) are shown) during the study period November 2002 to March 2003. 49

Table 4.2. Total abundance, number of taxa and Shannon Wiener Diversity Index during the study period within the middle and upper reaches of the Lourens River estuary. 53

Table 5.1 General water and sediment quality measurement ($n=2$) taken at the beginning of the experiment and after 96h. 67

Table 5.2 Measured chlorpyrifos concentrations in spiked sediment and overlying water in the various experimental treatments. 68

Table 5.3. Three factorial analysis of variance for the effects of adaptation to varying salinities (adapt), salinity (sal) and CPF exposure (cpf) on survival of female and male *Mesochra parva* (squareroot arcsin transformed) [a] 69

Table 6.1 Mean (± standard error) physicochemical parameters measured in the Rooiels River and Lourens River estuary from September 2001 to March 2003 (n=44) (Stars indicating the significant difference between the two estuaries (* $p < 0.05$; ** $p < 0.01$; *** $p < 0.001$)) 88

Table 6.2 Mean (± standard error) particle bound pesticide concentrations (μg/kg) in the Lourens and Rooiels River estuary and 90th percentile from September to March 2003 (n=27). (Asterisk indicating the significant difference between the two estuaries (* $p < 0.05$; ** $p < 0.01$)) 89

Table 6.3 Mean (± standard error) number of taxa, total abundance and Shannon Wiener Diversity Index in the Lourens and Rooiels River estuary and 90th percentile from September to March 2003 (n=27). (Asterisks*** indicating the significant difference between the two estuaries $p < 0.001$)) 93

List of Figures

Figure 2.1 False Bay and the location of the two studied temporarily open estuaries. 16

Figure 2.2 Direction of current during summer and winter within False Bay (Taljaard *et al.*, 2000) 17

Figure 2.3 Lourens River catchment (Taljaard *et al.*, 2000) 18

Figure 2.4 Site selection within the Lourens River estuary (Google Earth, Map Data © 2008; © 2008 Europa Technologies) 18

Figure 2.5 Rooiels River catchment (Taljaard *et al.*, 2000) 19

Figure 2.6 Site selection within the Rooiels River estuary (Google Earth, Map Data © 2008; © 2008 Europa Technologies) 20

Figure 3.1 Seasonal rainfall and air temperature during the study period from 2002 to 2003 (data obtained from the Verelegen weather station at 80m altitude). 24

Figure 4.1 Daily rainfall (obtained from Verlegen weather station at 80m altitude) and flow data (obtained from DWAF gauging station G2H029A01) during the study period from November 2002 to March 2003. (The dotted line indicates the threshold of 10mm runoff above which runoff is assumed to take place; arrows indicating two runoff events above 10mm) 43

Figure 4.2 Diagram showing the first component of the PRC of the differences in measured physico-chemical parameters between the middle and upper reaches. Sixty-three percent of the total variation in physico-chemical parameter composition could be attributed to differences between sampling dates, the other 37% to differences in physico-chemical parameter composition between the upstream and downstream sampling site; 47% of the latter is displayed in the diagram. The parameter weights shown in the right part of the diagram represent the affinity of each parameter with the response shown in the diagram. 46

Figure 4.3 Diagram showing the first component of the PRC of the differences in measured pesticide concentrations between the middle and upper reaches. Sixty-five percent of the total variation in pesticide composition could be attributed to differences between sampling dates, the other 35% to differences in pesticide composition between the upstream and downstream sampling site; 65% of the latter is displayed in the diagram. The pesticide weights shown in the right part of the diagram represent the affinity of each pesticide with the response shown in the diagram. 47

Figure 4.4 Diagram showing the first component of the PRC of the differences in species composition between the middle and upper reaches. Sixty-nine percent of the total variation in species composition could be attributed to differences between sampling dates, the other 31% to differences in species composition between the upstream and downstream sampling site. 36% of the latter is displayed in the diagram. The species weights shown in the right part of the diagram represent the affinity of each species with the response shown in the diagram. For the sake of clarity, only species with a species weight larger than 0.25 or smaller than -0.25 are shown. 48

Figure 4.6. Response of benthic taxa to the first runoff events within the upper and the middle reaches of the Lourens River estuary during the study period from November 2002 to March 2003. 54

Figure 4.7. Response of benthic taxa to the second runoff events within the upper and the middle reaches of the Lourens River estuary during the study period from November 2002 to March 2003. 55

Figure 5.1 Measured salinity fluctuation during the adaptation period of 28 days. 64

Figure 5.2 Illustration of the full factorial experimental design to evaluate the combined influence of adaptation to fluctuating salinity, salinity decrease and chlorpyrifos exposure in a 96-hour sediment toxicity test with *Mesochra parva* 64

Figure 5.3. Survival rates of male and female *Mesochra parva* in relation to the experimental treatments. Different letters indicate significant differences due to various combinations of experimental variables (Tukey's HSD test, $P < 0.05$), asterisks denote significant differences in survival of male and female organisms after the same treatment (Student's t-test, $P < 0.05$). Each column represents the mean of six replicates +/- standard deviation. 71

Figure 5.4. Measurements of salinity at the bottom (a), particle associated chlorpyrifos concentration (b) and *Mesochra parva* abundance (c) measured in the lower reaches of the Lourens River estuary between October 2001 and March 2003. Arrows in (a) indicate runoff events with amount of rainfall (a = 3 mm, b = 46.2 mm, c = 2.6 mm, d = 5.2, e = 51.8 mm). 72

Figure 6.1 False Bay and the location of Lourens River and Rooiels River estuary 82

Figure 6.2. Monthly rainfall and temperature data (obtained from Verlegen weather station at 80m altitude) during the study period from September 2001 to March 2003. (The dotted lines indicated the spraying season of pesticides). 84

Figure 6.3. Monthly composition of particle bound pesticide concentrations detected at Lourens and Rooiels River estuary from September 2001 to March 2003. Asterisks indicate that no samples were taken. 90

Figure 6.4. Diagram showing the first component of the PRC of the differences in measured environmental variables and pesticides between Lourens and Rooiels River estuary. Forty-two percent of the total variation could be attributed to differences between sampling dates, the other 58% to differences between the estuaries; 43% of the latter is displayed in the diagram. The parameter weights shown in the right part of the diagram represent the affinity of each parameter with the response shown in the diagram. For clarity, only parameters with a weight > 0.5 or < -0.5 are shown. 91

Figure 6.5. Diagram showing the first component of the PRC of the differences in species composition between the Rooiels and Lourens rivers. Fifty-two percent of the total variation in species composition could be attributed to differences between sampling dates, the other 48% to differences in species composition between the estuaries; 44% of the latter is displayed in the diagram. The taxa weights shown in the right part of the diagram represent the affinity of each taxa with the response shown in the diagram. For the sake of clarity, only species with a weight larger than 0. 5 or smaller than -0. 5 are shown. 92

Figure 6.6. RDA triplot resulting from an analysis using the environmental variables and pesticide concentrations that explain a significant part of the seasonal variation in the species composition in the Rooiels River estuary as explanatory variables and meiofauna taxa composition as taxa data. The two significant variables explain 14% of the seasonal variation in species composition, of which 71% is displayed on the first axis and another 29% on the second one. For clarity only the 18 most important taxa out of 33 are shown..94

Figure 6.7. RDA triplot resulting from an analysis using the environmental variables and pesticide concentrations that explain a significant part of the seasonal variation in the species composition in the Lourens River estuary as explanatory variables and meiofauna taxa composition as taxa data. The six significant variables explain 43% of the seasonal variation in species composition, of which 69% is displayed on the first axis and another 12% on the second one. For clarity only the 20 most important taxa out of 33 are shown. 95

SUMMARY

The research described in this thesis was designed to yield information on the impact of particle-bound pesticides on organisms living in the interface between sediment and water column in a temporarily open estuary (TOCEs). It was hypothesized that natural variables such as salinity and temperature and anthropogenic stressors such as particle-bound pesticides contribute to the variability of the system. A multiple line of evidence approach is necessary due to the variability in sediment type, contaminant distribution and spatial and temporal variability within the ecosystem in particular within TOCEs. Accordingly, sediment contamination in comparison with guideline values including Equilibrium Partitioning (EqP) modelling (Chapter 3), field investigations of sediment toxicity and benthic community structure (Chapter 4, 6) and laboratory spiked sediment toxicity tests (Chapter 5) remains the most accurate way to account for contaminant availability and effects.

The first aim of this thesis was to identify which particle-bound pesticides are important to the contamination of the Lourens River estuary (Western Cape, South Africa), taking into account their environmental concentrations, physico-chemical and toxicological properties (Exposure assessment).

The use of EqP theory, Species Sensitivity Distribution and the comparison with threshold values from various international water quality guidelines allowed the authors to evaluate a potential risk of pesticides towards marine and freshwater communities. Particle associated pesticides (chlorpyrifos, prothiofos, cypermethrin, fenvalerate, endosulfan and p,p-DDE) and physicochemical parameters (salinity, temperature, flow, and total organic content) were measured in the Lourens River estuary twice a month over a period of two years and the EqP theory was applied to calculate concentrations of pesticides in the water. Highest 90th percentile concentrations of pesticides associated with suspended particles were 33.96 µg/kg for prothiofos, 19.56 µg/kg for chlorpyrifos and 18.61 µg/kg for endosulfan. The calculated concentrations in the water were 0.15, 0.24 and 1.25 µg/l for prothiofos, chlorpyrifos and endosulfan, respectively.

Species Sensitivity Distribution was used to determine the predicted hazardous concentration (HC5) for freshwater and marine communities and indicated a higher higher sensitivity for marine organisms. The calculated concentrations in the water exceeded all threshold values suggested by international water guidelines. Chlorpyrifos and endosulfan posed the highest risk towards freshwater and marine communities in the Lourens River estuary due to the lowest Toxicity Exposure Ratio (lowest threshold value in combination with highest exposure concentration). No sufficient toxicity data and threshold values were found for prothiofos, thus it was not possible to assess a possible risk for this pesticide towards the aquatic environment. The spring seasons were characterised by the highest pesticide exposure due to the slow flow and frequent pesticide application and by the highest pesticide toxicity due to the increase in salinity and in abundance of more sensitive marine organisms.

The second aim was to identify spatial and temporal variations in particle bound pesticide contamination, natural environmental variables and benthic community structure (effect assessment). The study focused on the effect of particle associated pesticides on the dynamics of the benthic community by comparing two runoff events, differing in their change in pesticide concentration and environmental variables. Additionally different methods of community description were explored and compared with each other in order to determine which environmental variables, particle bound pesticides and which taxa contribute to spatial and temporal differences, including community indices (like species richness and Shannon Diversity Index) and a multivariate approach, namely Principle Response Curve. The two chosen sites were situated within the upper and middle reaches of the estuary and differ significantly mainly in salinity ($p=0.001$), flow ($p=0.5$), temperature ($p<0.001$) and total organic carbon in the sediment ($p<0.001$). Generally higher particle bound pesticides were found in the upper reaches.

The first runoff event was characterised by an increase in pesticides (chlorpyrifos, endosulfan and cypermethrin) and hardly any change in natural environmental variables, whereas the second runoff event was characterised by no increase in pesticide but a significant change in natural environmental variables like salinity, temperature and flow. The use of the Principal Response Curve showed a clear difference between the two sites but also a high temporal variability in environmental variables (explaining 63% of variation), particle bound pesticides (65%) and benthic community (69%). The most evident spatial difference in community structure was shown after the first runoff event, whereas no response was shown after the second runoff event. The variables which explained most of the spatial differences are total Organic carbon, salinity, phosphate and endosulfan concentrations. The species contributing most to the differences between the sites are the estuarine harpacticoids *Mesochra* and *Canthocamptus* (lower abundance at the upper reaches) and the freshwater species *Dunhevedia* and *Thermocyclops* (higher abundance within upper reaches). Therefore Principal Response curve was shown to be a useful tool for explaining temporal and spatial variability within a temporarily open estuary using biomonitoring data.

The third aim was to test the hypothesis: "does adaptation to fluctuating salinities lead to enhanced survival of the harpacticoid copepod *Mesochra parva* when exposed to a combination of particle associated chlorpyrifos exposure and hypo-osmotic stress during a 96 h sediment toxicity test?" The chlorpyrifos exposure concentrations of 5.89-5.38 µg/kg and the salinity decrease from 15 ppt to 3 ppt were based on conditions observed in the temporarily open Lourens River estuary, South Africa, in order to simulate changes during a runoff event. Results of the three-factorial ANOVA showed that pre-adaptation to varying salinities ($p=0.02$; $p=0.001$), salinity decrease ($p=0.035$; $p<0.001$) and CPF exposure ($p<0.001$; $p<0.001$), all had a significant negative impact on the survival rate of female and male *M. parva*, with a higher sensitivity of males specimens. The significant two-way interaction of salinity x adaptation for females and males ($p=0.021$; $p<0.001$), indicate that adaptation to fluctuating salinities was beneficial for male and female copepods, but the hypothesis of a three-way interaction was not supported. However a trend indicated a lower survival rate of non-adapted females and males exposed to CPF and hypo-osomotic stress ($38 \pm 17\%$; $0 \pm 0\%$), compared to pre-adapted organisms ($59 \pm 6.6\%$; $8.9 \pm 10\%$), which requires further elucidation.

The last aim was to identify the driving environmental variables (including natural and anthropogenic stressors) in a "natural" (Rooiels River) compared to a "disturbed" (Lourens River) estuary and to identify if and how these variables change the benthic community structure in both estuaries.

Due to the natural much smaller catchment of the Rooiels River estuary, many environmental variables were significantly different ($p< 0.001$) from the variables in the Lourens River estuary, like salinity, temperature, pH, total suspended solids, nitrate and depth. No pesticide concentrations were expected in the Rooiels River estuary due to no associated agricultural catchment. However, chlorpyrifos (8.9 µg/kg), prothiofos (22.0 µg/kg) and cypermethrin concentrations (0.42 µg/kg) were detected frequently with highest concentrations during the summer months. The use of the Principal Response

Curve Analysis showed that temporal variability between sampling dates explained 42% of the variance in environmental variables and pesticide concentrations and spatial variability between the two estuaries explained 58%. Variables contributing mostly to the difference were higher concentrations of endosulfan, p,p-DDE and nitrate concentrations in the Lourens River estuary and larger grain size and higher salinity at the bottom in the Rooiels River estuary. In general the meiofauna community in the Rooiels River estuary showed a significantly higher number of taxa ($p < 0.001$), a significant higher Shannon Wiener Diversity Index ($p < 0.001$) and a generally lower abundance with less variability than in the Lourens River estuary. The differences were mostly explained by a higher abundance of *Cypretta* and *Darcythompsonia* in the Rooiels River estuary and a higher abundance of *Thermocyclops* and *Canthocamptus* in the Lourens River estuary.

The variables explaining a significantly part (14%) of the variance in meiofauna abundance in the Rooiels River estuary were salinity and temperature. The Redundancy Analysis indicated that most of the taxa were shifted towards high salinity and temperatures. Taxa like *Upogebia*, *Nereis*, *Uroma* and nematodes were clearly positively correlated to salinity and temperature. Thus in general this estuary was dominated by estuarine (euryhaline) and marine taxa. The variables explaining a significant part of the variance in the dataset (43%) within the Louren River estuary, however, were salinity and temperature on the one hand and chlorpyrifos, nitrate and flow on the other hand. Although the Rooiels River estuary showed some concentrations of particle bound pesticides, none of them influenced the community structure in flocculent layer significantly.

Data produced in this research thus provide important information to understand the impact of pesticides and its interaction with natural variables in a temporarily open estuary. To summarise, this study indicated, by the use of the multi-evidence approach, that the pesticides endosulfan and chlorpyrifos posed a risk towards benthic organisms in a temporarily open estuary in particular during spring season. Furthermore an important link between pesticide exposure/ toxicity and salinity was identified, which has important implications for the management of temporarily open estuaries.

ZUSAMMENFASSUNG

Die in in dieser Studie beschriebene Forschungsarbeit hatte primär das Ziel, den Einfluss partikelgebundener Pestizide auf Organismen in der flockenartigen Schicht zwischen Sediment und Wasser in einem temporär offenen Ästuar zu erfassen. Dabei wurde die Hypothese aufgestellt, dass natürliche Variablen wie Salzgehalt und Temperatur sowie anthropogene Stressfaktoren wie partikelgebundene Pestizide zur Variabilität des Systems beitragen. Die Variabilität der Sedimentzusammensetzung, der Verteilung der Kontamination, sowie der räumlichen und zeitlichen Unterschiede im Ökosystem, speziell in temporär offenen Ästuaren, macht hier einen kompositionellen Ansatz notwendig. Demenstsprechend sollten folgende Aspekte einbezogen werden, um die Verfügbarkeit und den Einfluss von Schadstoffen möglichst akkurat untersuchen zu können: Der Vergleich von Schadstoffkonzentrationen im Sediment mit Richtwerten einschließlich Gleichgewichts-Verteilungs-Modellierung (Kapitel 3), die Einschätzung der Sedimenttoxizität im Freiland (Kapitel 4, 6) und beimpfte Sedimenttoxizitätstests im Labor.

Das erste Ziel dieser Arbeit war die Identifikation der partikelgebundenen Pestizide, die zur Kontamination des Lourens-River-Ästuars beitragen, unter Berücksichtigung ihrer Umweltkonzentrationen sowie physikalisch-chemischer und toxischer Eigenschaften (Expositionsanalyse). Durch die Anwendung der Gleichgewichts-Verteilungstheorie, Arten-Empfindlichkeitsverteilung und dem Vergleich mit Grenzwerten verschiedener internationaler Wasserqualitätsrichtlinien wurde das potentielle Risiko von Pestiziden gegenüber marinen und Süßwassergemeinschaften evaluiert. Über einen Zeitraum von zwei Jahren wurden zweimal monatlich die Konzentrationen partikelgebundener Pestizide (Chlopryrifos, Prothiofos, Cypermethrin, Fenvalerate, Endosulfan und p,p-DDE) und physikalisch-chemische Parameter (wie z.B. Salzgehalt, Temperatur, Fließgeschwindigkeit und organischer Gehalt im Sediment) im Lourens-River-Ästuar gemessen, wobei die Pestizidkonzentrationen im Wasser unter Anwendung der Equilibriums-Verteilungs-Theorie berechnet wurden. Die höchste 90. Perzentilkonzentration partikelgebundener Pestizide belief sich auf 33,96 µg/kg für Prothiofos, 19,56 µg/kg für Chlopryrifos und 18,61 µg/kg für Endosulfan. Die berechneten Konzentrationen im Wasser betrugen 0,15, 0,24 und 1,25 µg/l für Prothiofos, Chlorpyrifos und Endsulfan. Die Rangordnung aquatischer Invertebraten hinsichtlich ihrer Empfindlichkeit gegenüber Pestiziden (Arten-Empfindlichkeitsverteilung) wurde angewendet, um die Konzentrationen, bei denen eine Wirkung gegenüber Süßwasser- und marinen Artengemeinschaften zu erwarten ist, berechnen zu können. Dabei wurde eine höhere Empfindlichkeit für marine Organismen festgestellt. Die berechneten Konzentrationen im Wasser überschritten die Grenzwerte aller internationalen Richtlinien. Aufgrund des niedrigen Toxizitäts-Expositionsverhältnisses (Niedrigster Grenzwert im Verhältnis zu höchster Exposition) stellen Chlorpyrifos und Endosulfan das höchste Risiko für marine und Süßwasser-Artengemeinschaften im Lourens-River-Ästuar dar. Da für Prothiofos nur unzureichende Toxizitätsdaten und Grenzwerte vorliegen, konnte das Risiko dieses Pestizides nicht hinreichend eingeschätzt werden. Im Frühjahr wurden zum einen die höchsten Pestizidkonzentrationen gemessen, was auf die geringere Fließgeschwindigkeit sowie regelmäßige Pestizidanwendung zurückzuführen ist; zum anderen wurde in diesem Zeitraum aufgrund des steigenden Salzgehaltes sowie dem vermehrten Vorkommen empfindlicher mariner Arten die höchste Toxizität festgestellt.

Das zweite Ziel war die Identifikation räumlicher und temporärer Variabilität partikelgebundener Pestizide, natürlicher Umweltvariablen und benthischer Artenzusammensetzung (Wirkungsanalyse). Die Studie konzentrierte sich auf die Auswirkungen partikelgebundener Pestizide auf die Dynamik benthischer Artengemeinschaften, indem die Wirkung zweier Oberflächenabflussereignisse, die sich in den Änderungen von Pestizidkonzentrationen sowie diversen Umweltvariablen unterschieden, verglichen wurde. Zusätzlich wurden verschiedene Methoden der Artenbeschreibung, wie Artenreichtum und *Shannon Diversity Index*, sowie eine multivariate statistische Methode (*Principal Response Curve*) getestet und miteinander verglichen, um festzustellen, welche Umweltvariablen und Taxa, sowie partikelgebundene Pestizide zu den räumlichen und zeitlichen Unterschieden beitragen. Die zwei Probestellen in dieser Studie befanden sich im oberen und mittleren Bereich des Ästuars und unterschieden sich signifikant hauptsächlich in den Bereichen Salzgehalt ($p=0,001$), Fließgeschwindigkeit ($p=0,5$), Temperatur ($p<0,001$) sowie organischem Anteil im Sediment ($p<0,001$). Im Allgemeinen wurden im oberen Bereich des Ästuars höhere Konzentrationen an partikelgebundenen Pestiziden gemessen.
Das erste Oberflächenabflussereignis verursachte einen Eintrag von Pestiziden (Chlorpyrifos, Endosulfan und Cypermethrin), ohne jedoch größere Veränderungen in

natürlichen Umweltvariablen hervorzurufen. Das zweite Oberflächenabflussereignis hingegen verursachte keinerlei Eintrag von Pestziden, jedoch signifikante Veränderungen in natürlichen Umweltvariablen wie Salzgehalt, Temperatur und Fließgeschwindigkeit.

Die Anwendung der multivariaten Statistik (*Principal Response Curve*) zeigte nicht nur einen eindeutigen räumlichen Unterschied zwischen den beiden Probestellen sondern auch eine hohe zeitliche Variabilität von natürlichen Umweltvariablen (die 63% der Variation erklärt), partikelgebundener Pestizide (65%) und benthischer Artengemeinschaft (69%). Der groesste raeumliche Unterschied wurde durch das erste Oberflaechenabflussereignis hervorgerufen, wohingegen das zweite Ereignis keine Aenderungen hervorrufte. Die Variablen, die den räumlichen Unterschied am deutlichsten erklärten, waren der organische Gehalt im Sediment, Salzgehalt, sowie Phosphat- und Endosulfankonzentrationen. Die Arten, die am meisten zum Unterschied zwischen den Probestellen beitrugen waren die ästuaren harpacticoden Arten *Mesochra* und *Canthocamptus* (mit geringerer Abundanz im oberen Bereich des Ästuars) und die Süßwasserarten *Dunhevedia* und *Thermocyclops* (mit höherer Abundanz im oberen Bereich des Ästuars). Die Studie zeigte daher, dass die multivariate statistische Methode *Principal Response Curve* erfolgreich die räumliche und zeitliche Variabilität innerhalb des Ästuars anhand von Freilanderfassungen erklären konnte.

Das dritte Ziel der Studie war es, die Hypothese zu testen, ob Anpassung an schwankende Salzgehalte die Überlebensrate von einem harpacticoiden Copepoden *Mesochra parva* fördert, bei gleichzeitiger Aussetzung von Chlorpyrifos und hypoosmotischem Stress während eines 96-stündigen Sedimenttoxizitätstests. Die gemessenen Chlorpyrifoskonzentrationen im Experiment von 5,89-5,38 µg/kg und die Salzgehaltsverminderung von 15 ppt auf 3 ppt sind freilandrelevante Bedingungen und simulierten ein Oberflächenabflussereignis im Lourens-River-Ästuar. Ergebnisse der dreifaktoriellen ANOVA zeigten, dass die vohrige Anpassung an verschiedene Salzgehalte ($p=0{,}02$; $p=0{,}001$), Salzgehaltsverminderung ($p=0{,}035$; $p<0{,}001$) und Exposition zu Chlorpyrifos ($p<0{,}001$; $p<0{,}001$) die Überlebensrate von weiblichen und männlichen *M.-parva*-Organismen beeinflussten, wobei die Männchen eine höhere Empfindlichkeit zeigten. Eine statistisch signifikante Interaktion fuer die Förderung der Überlebensrate von Männchen und Weibchen wurde für die Kombination Salzgehalt x Anpassung festgestellt; die Hypothese einer Interaktion zwischen allen drei experimentellen Faktoren wurde jedoch nicht bestätigt. Allerdings wird eine Tendenz angedeutet, dass unangepasste Weibchen und Männchen, die gleichzeitig Chorpyrifos und hypoosmotischen Stress ausgesetzt sind, eine niedrigere Überlebensrate aufzeigen (38 ± 17%; 0 ± 0%), als angepasste Organismen (59 ± 6,6%; 8,9 ± 10%), was eine nähere Untersuchung erforderlich macht.

Das letzte Ziel der Studie beinhaltete die Identifikation jener Umweltvariablen (einschließlich natürlicher und anthropogener Stressoren) die ein Ästuar beinflussen. Zu diesem Zweck wurde ein natürliches Ästuar (Rooiels River) mit einem anthropogen beeinflussten Ästuar (Lourens River) verglichen. Zudem sollte aufgezeigt werden, ob und inwiefern sich diese Variablen auf die benthische Artengemeinschaft in beiden Ästuaren auswirken. Aufgrund des natürlichen, viel kleineren Einzugsgebiets des Rooiels-River-Ästuars unterschieden sich zahlreiche Variablen wie Salzgehalt, Temperatur, Trübung, Nitratkonzentration und Wassertiefe von denen im Lourens-River-Ästuar. Obwohl keinerle Pestizidkonzentrationen im Rooiels-River-Ästuar vermutet wurden, da keinerlei Landwirtschaft im Einzuggebiet festzustellen war, wurden regelmäßig Konzentrationen von Chlorpyrifos (8,9 µg/kg), Prothiofos (22,0 µg/kg) und Cypermethrin (0,42 µg/kg) festgestellt, wobei die Werte während der Sommermonate am höchsten waren. Die Anwendung der multivariaten Statistik (*Principal Response Curve*) zeigte, dass die zeitliche Variabilität zwischen den Probedaten 42% der Varianz der Umweltparameter und Pestizidkonzentrationen erklärte und dass die übrigen 58% durch Pestizidkonzentrationen und die räumliche Varibilität zwischen den beiden Ästuaren erklärt werden konnten. Die Parameter, die am meisten zu den Unterschieden beitrugen, waren höhere Endosulfan-, p,p-DDE- und Nitratkonzentrationen im Lourens-River-Ästuar sowie die größere Korngröße des Sediments und höherer Salzgehalt im Rooiels-River-Ästuar. Im Allgemeinen zeichnete sich das Rooiels-River-Ästuar im Vergleich zum Lourens-River-Ästuar durch eine signifikant höhere benthische Artenanzahl (p<0,001), einen signifikant höheren *Shannon Wiener Diversity Index* (p<0,001) und einer generell niedrigeren Abundanz mit geringerer Variabilität aus. Die Unterschiede zwischen der beiden Ästuaren konnten dabei größtenteils durch eine höhere Abundanz von *Cypretta* und *Darcythompsonia* im Rooiels-River-Ästuar und einer höheren Abundanz von *Thermocyclops* und *Canthocamptus* im Lourens-River-Ästuar erklärt werden. Salzgehalt und Temperatur sind die einzigen Parameter, die einen signifikanten Teil der Varianz der Artengemeinschaft im Rooiels-River-Ästuar erklärten

(14%). Die Redundanzanalyse zeigte auf, dass die meisten Arten (wie *Upogebia*, *Nereis*, *Uroma*) mit steigendem Salzgehalt und steigender Temperatur positiv beeinflusst wurden. Aufgrunddessen herrschen im Rooiels-River-Ästuar euryhaline und marine Arten vor. Im Lourens-River-Ästuar hingegen erklären zum einen Salzgehalt und Temperatur und zum anderen Chlopyrifos, Nitrat und Fließgeschwindigkeit die Varianz der Artengemeinschaft zu 43%. Obwohl im Rooiels-River-Ästuar Pestizidkonzentrationen nachgewiesen wurden, wurde die Artengemeinschaft dort nicht signifikant beeinflusst.

Die Ergebnisse dieser Arbeit sind somit für das Verständis der Wirkung von Pestiziden und ihrer Interaktion mit natürlichen Parametern in einem temporär offenen Ästuar von großer Bedeutung. Zusammenfassend zeigt diese Arbeit mit ihrem kompositionellen Ansatz auf, dass Pestizide wie Endosulfan und Chlopyrifos ein Risiko für die benthische Artengemeinschaft in einem temporär offenen Ästuar darstellen, insbesondere während des Frühjahrs. Desweiteren wurde nachgewiesen, dass Exposition und Toxizität von Pestiziden eng mit dem Salzgehalt in temporär offenen Ästuaren verknüpft sind. Diese Erkenntnis hat wichtige Folgen für den Schutz von temporär offenen Ästuaren.

CHAPTER 1

GENERAL INTRODUCTION

1.1 Introduction

Agricultural practice in South Africa is becoming increasingly dependent on the widespread use of toxic pesticide chemicals for crop protection, growth regulation and seed treatments to enhance productivity (London and Myers, 1995). Thus, South Africa forms 60% of the pesticide market in Africa (Dinham, 1993). The pesticide use under African conditions in many instances differs drastically from the intended agricultural conditions of use in the countries where they have been developed (Bouwman, 2004) and as a consequence, has led to contamination of the environment. This includes inadequate pesticide management, control, funding, enforcement frameworks and disposal facilities (Naidoo and Buckley, 2003). The contamination of surface waters is especially a cause of recent public concern (Heath and Claassen, 1999; Bollmohr *et al.*, 2009a).

In recent years, various investigations have been conducted in South Africa to determine the concentrations of a number of dissolved and particle-bound pesticides (Schulz, 2001; Schulz *et al.*, 2001; Dabrowski *et al.*, 2002). Of the studies that have been carried out, very few have attempted to establish a direct link between pesticide concentrations detected in the environment and effects (Thiere and Schulz, 2004, Bollmohr and Schulz, 2009).

Information on the concentrations of pesticides and their effects in estuaries is sparse internationally (Steen *et al.*, 1999; Steen *et al.*, 2001; Bondarenko *et al.*, 2004; Zulin *et al.*, 2002) and there exists a complete knowledge gap in South Africa in that regard. Such investigations are necessary to understand the holistic risk that pesticides pose along the entire river stretch, including estuaries with different environmental variability regimes and different community structures. The gap is further extended by most of the studies being focused on the exposure and effect of dissolved pesticides, whereas only a few studies looked at particle-bound pesticides (Bergamaschi *et al.*, 1999; MacDonald *et al.*, 1996).

Suspended particles, originating from runoff in the upper catchment (Schulz, 2001) and transported into estuaries, can accumulate during low flow season, characterised by very little exchange with the sea. These particles entering the river have been shown to be associated with peak high concentrations of pesticides like chlorpyrifos (924 µg/kg), total endosulfan (12,082 µg/kg) and prothiofos (980 µg/kg) (Schulz, 2001). Thus particle-bound pesticides are of high importance in estuaries. The bioavailability might be lower, however than dissolved pesticides (Chapman and Wang, 2001) and the effect assessment is a necessary step to understand their ecological risk. The approach used in this thesis to assess sediment quality in an estuarine system by integrating measures of chemical contamination concentration, Equilibrium partitioning (EqP) modeling, sediment toxicity and various measures of benthic community structure as is suggested by Chapman (1996). By considering several indicators of endpoints, the study provides a multiple-evidence approach to assess the impact of particle-bound pesticides as it is used by several legislators to derive numerical sediment quality guidelines (Chapman, 1996; Burton et al., 2002). However this analysis is only possible if the natural variability of the system is understood (Kibirige and Perissinotto, 2003) by including the extensive measurement of several natural variables (e.g. salinity, grain size, total organic carbon, temperature, etc).

1.2 Temporarily open estuaries

South African estuaries have been separated into five main categories, permanently open estuaries, temporarily open/closed estuaries (TOCEs), estuarine bays, estuarine lakes and river mouths (Whitfield, 1992). Temporarily open/closed estuaries are the dominant estuary type in the country, comprising 73% of all 258 estuaries (Whitfield, 1992).

Two processes were defined as important "services" provided by estuaries: (A) provision of nursery habitat for marine species and therefore an important economic contribution to inshore fisheries and (B) contribution to marine ecosystem productivity by transportation of sediment and nutrients into the marine zone (Turpie, 2002). Furthermore, estuaries provide immense recreational services and important resources for subsistence to communities living close to shore.

Similar systems, sometimes referred to as "blind", "intermittently open", or "seasonally open" estuaries", are only found in Australia, on the west coast of the USA, South America and India (e.g. Ranasinghe and Pattiaratchi, 1999). Temporally open/closed estuaries (TOCEs) do not have a permanently open link to the sea. During the dry season and under low flow conditions (20% of MAR) they are closed off from the sea by a sandbar. During this period the residence time (the time it takes to exchange all the water in a system with external water (Reddering, 1988)) increases, which may result in a higher influence of environmental degradation on the resident community compared to the permanent open estuaries. The estuary is stratified in salinity, temperature and oxygen and may become hypersaline due to evaporation during the summer. Following periods of high rainfall and freshwater runoff, the water level in the estuary may rise until it exceeds the height of the sandbar at the mouth (Wooldridge & Callahan, 2000) and breaching occurs. The opening of the estuary coincides with a rapid drop in the water level and river conditions may briefly dominate.

Depending on the climatic conditions and rainfall patterns in the catchment areas, closure periods may vary naturally from days to months or even years. This extremely dynamic situation has important implications for the variability of physical, chemical and biotic parameters of TOCEs and ultimately for their ecological functioning (Kibirige & Perissinotto, 2003). Some studies have focussed on seasonal fluctuations of meiofauna and zooplankton in South African estuaries in relation to the distribution of physical factors (Dye, 1983; Nozais *et al.*, 2005), state of the estuary mouth (Kibirige & Perissinotto, 2003) and food web interactions (Perissinotto *et al.*, 2000) in order to understand the impact of temporal variability on the ecological functioning of the system. In addition to general assessments of health of estuaries based on freshwater availability, some work has been carried out on the health of certain biotic and abiotic components of estuaries (Heydorn, 1982). Harrison *et al.* (2000) present an assessment of the health of all South African estuaries in terms of ichthyofaunal diversity, general water quality and aesthetics, and Coetzee *et al.* (1997) and Colloty *et al.* (2000) have classified selected estuaries in terms of their botanical integrity. However, no study has been performed to understand the dynamic of natural variables (mainly salinity) in combination with anthropogenic variables and their effects on the dynamic of community structures in such highly variable systems like TOCEs.

1.3 Anthropogenic disturbances/ catchment importance

Changes to the flow into estuaries can have a significant impact on mouth dynamics and therefore on the overall functioning of estuarine systems. Retention or abstraction of freshwater for industrial, agricultural and domestic purposes has led to a reduction of both the frequency and duration of mouth-opening (Reddering, 1988). Estuaries with a high percentage of agriculture in their catchment are especially affected, since just over half of all available freshwater is consumed by agricultural irrigation systems (Clarke, 1993).

Since TOCEs have normally small catchments and since estuaries as water end-users of the entire catchment reflect the land use of the entire catchment, they exhibit a particularly high sensitivity towards land use changes (Turpie et al., 2002). Furthermore, they are important zones to trap natural and anthropogenic stressors due to decrease in flow and increase in salinity before discharging into the sea. Deforestation, wetlands destruction and overgrazing in catchment areas are a major cause of soil erosion in South Africa. Poor agricultural practices in the catchment can lead to increased nutrient, pesticide and sediment loads in rivers and their estuaries while domestic and industrial waste discharge also introduce pollutants (e.g. trace metals) into these systems. Agricultural impacts on water quality and living resources in the coastal zone have been well documented internationally over the past half-century, with examples ranging from sedimentation (Valette-Silver et al., 1986), eutrophication (Carpenter et al., 1998), to a few examples of pesticide toxicity (e.g. Clark et al., 1989, 1993).

1.4 Exposure assessment

The exposure assessment forms the basis of any risk assessment and forms the part of the characterization of the multiple-evidence- approach.

With respect to agricultural and rural environments, nonpoint-source pollution is generally regarded as the most important source of contaminants in surface waters (Loague et al., 1998), and runoff and spray-drift in particular, have been identified as the most important sources of pesticides (Flury, 1996). Pesticides can be found in two forms in surface runoff; in the soluble form (dissolved in runoff water) and in the eroded form (sorbed to suspended solids). Hydrophobic organic chemicals, including pesticides with a high KOC, tend to adsorb to suspended solids (Voice and Weber, 1983; Ingersoll et al., 1995). These particle-bound pesticides are of particular importance during runoff events (after heavy rainfalls, Ingersoll et al., 1995) and may accumulate in sediments of TOCEs.

Contaminated sediments pose a major environmental hazard primarily because the sediments act as a major long-term storage and source of toxic chemicals discharged into surface waters (Burton, 1991). The South African Department of Water Affairs and Forestry acknowledged the fact that many toxicants being monitored in the National Toxicity Monitoring Programme (DWAF, 2006) are bound to the sediment than being dissolved in the water (Bollmohr et al., 2009a).

Sediments are composed of heterogeneous mixtures of detritus, organic and inorganic particles that settle at the bottom of a body of water (Power and Chapman, 1992). The inorganic particles include rock and shell fragments and mineral grains and the organic contents are usually a small fraction of the total sediment volume (Power and Chapman, 1992). However, organic matter is an important food source for benthic organisms and it has a major role in regulating the sorption and bioavailability of many contaminants (Reuber et al., 1987; Grathwohl, 1990). Particle-bound pesticides might be of higher importance in estuaries than in rivers, since estuarine community structures consist of more particle feeders than fresh water communities, with a large fraction of organisms situated at the interface between water and sediment (flocculent layer), directly exposed to particle-bound contaminants (Chandler et al., 1997). Interstitial water fills the spaces between sediment particles and the partitioning of contaminants between sediment organic matter and interstitial water is an important process responsible for the fate, transport and bioavailability of hydrophobic contaminants (Ankley et al., 1994; Segstro et al., 1995).

The Equilibrium Partitioning (EqP) method (diToro et al., 1991) is used to estimate sediment toxicity from water quality criteria, when insufficient sediment toxicity data are available (van Beelen et al., 2003) and many authors used this approach to determine toxicity of particle-bound pesticides (Ankley et al., 1994; Green et al., 1996; Chandler and Green, 2001; Villa et al., 2003a) (Chapter 3). The EqP theory assumes that a chemical bound to sediment organic carbon is in thermodynamic equilibrium with the chemical dissolved in the interstitial water and the lipid components of the exposed organism (McCauley et al. 2000), depending on the organic carbon content in the sediment and the organic/carbon water partitioning coefficient of the chemical.

Most of the studies in estuaries focused on dissolved pesticide concentrations. Kuivila and Foe (1995) detected a high spatial and temporal variability of diazinon, methidathion and chlorpyrifos concentrations in the San Francisco estuary linked to rainfall events, of which diazinon posed the highest risk. Sujatha et al., (1999) detected a temporal and spatial variability of endosulfan and malathion concentrations in an Indian estuary, explained by changes in salinity, pH and sedimentation. Steen et al. (1999) compared the ecological risk of various pesticides related to different European estuaries and concluded that the sum of all pesticides exert a significant pressure on the aquatic ecosystem. However, they also identified a lack of toxicity data to be used in the ecological risk assessment. Another study by Steen et al. (2001) established a temporal link between pesticides (and their breakdown products) detected in the Scheldt estuary and the time of application. Bondarenko et al., (2004) determined a longer persistence of chlorpyrifos and diazinon in a coastal watershed in southern California, due to salinity and inhibition of microbial degradation. Zulin et al. (2002) detected 17 organophosphorus and 19 organochlorine pesticides in an estuary in China, and identified metamidophos, dichlorvos and dimethoate as the pesticides posing the highest risk. In a study by Villa et al., 2003a most of the analyzed 10 different pesticides in the water of a lagoon ecosystem were found to be below the detection limit, due to the very sporadic occurrence of pesticides in the water.

Particle-bound pesticides were analysed in only a few studies. A study by Bergamaschi et al. (1999) detected particle-bound pesticides more frequently than dissolved pesticides in San Francisco Bay, with concentrations ranging from 0.6 to 2.1 µg/kg for chlorpyrifos and 17.7 to 24.6 µg/kg for endosulfan, emphasizing the inclusion of particle-bound pesticides in monitoring programs. Within the National Status and Trend Monitoring Program, various bays in USA have been analyzed for different particle-bound pesticides (MacDonald et al., 1996) in order to derive sediment quality guidelines. However, the Lourens River estuary in South Africa, studied in this thesis, is known to

General Introduction

receive particle-bound pesticides like chlorpyrifos (924 µg/kg), total endosulfan (12,082 µg/kg) and prothiofos (980 µg/kg) from further upstream (Schulz, 2001).

The exposure assessment has been performed in order to understand the temporal (Chapter 3, 4, 6) and spatial (Chapter 4) variability of particle-bound pesticides and natural variables within an impacted estuary (Lourens River estuary) and in order to compare the variability with a natural estuary (Rooiels River estuary) (Chapter 6).

1.4 Effect assessment

In ecological risk assessment, endpoints and measures of effects can be defined at all levels of organization, ranging from the individual to the community level (Suter *et al.*, 1993).

A large proportion of the knowledge of effects of pesticides on estuarine and marine species have been based on standard toxicity tests that measure endpoints (e.g. mortality, immobility or reproduction) of single species exposed to a range of increasing toxicant concentrations. The calculated LC50 or EC50 values are often extrapolated to the field to provide an indication of the potential toxicity of contaminants measured in the field. The exposure duration in standardised test (e.g. static 96-h tests) is often criticised as not being field relevant, since organisms in a flowing system are typically exposed in pulses and not in continuous exposure (Reinert *et al.*, 2002). However, in temporarily open estuaries, in which particles bound pesticides can be accumulated and remain for a long time, organisms can be exposed over a long time. Various sediment toxicity tests have been performed in order to determine the acute toxicity of azinphos-methyl (Klosterhaus *et al.*, 2003), fenvalerate (Chandler *et al.*, 1994), endosulfan (Chandler and Scott, 1991) and chlorpyrifos (Green *et al.*, 1996) with a suggested interaction between salinity and chlorpyrifos exposure of meiobenthic harpacticoid copepods (Staton *et al.*, 2002).

Single-species toxicity tests do not, however, take into account interactions between- and within species and may thus not be adequate for extrapolating the impact at community or population level. Although a toxicant may directly affect species that are exposed and vulnerable to its mode of action, the change in community structure exposed to a toxicant is rather an indirect effect of toxicity on species interactions (Fleeger et al., 2003). For example, while slow-growing, stress-tolerant species may have had poor fitness relative to fast growing, stress-sensitive species before being exposed to a toxicant, their subsequent resistance promotes a better relative fitness compared to the fast growing species (Preston 2002).

Because of limitations associated with the use of standard test organisms, an alternative measure, namely the species sensitivity distribution (SSD), has been used in risk assessment (Forbes and Calow, 2002; Maltby et al., 2005). By testing a number of organisms, a distribution of sensitivity is obtained to assess the fraction of species affected in the environment. For example, a concentration of a chemical that is expected to affect 5% of the population in a system is statistically derived and denoted HC5 (for hazard concentration at 5% effect). A major limitation of this approach, however, is that extensive toxicological data are required for the great variety of species assemblages that exist in the field. Insufficient toxicity data often exist for marine organisms to estimate a marine HC5 (Wheeler et al., 2002). Marine species tend to be more sensitive to pesticide compounds than freshwater species, thus the extrapolation of toxicity towards freshwater taxa to toxicity towards marine taxa is unclear (Chapter 3; Wheeler et al., 2002). In addition, such species sensitivity distributions refer to one substance only, and do not consider interaction with natural variables or with other contaminants.

As a result, increased emphasis has been placed on developing more field-relevant toxicity tests so as to better understand the effects of pesticides on aquatic communities. Various freshwater experiments range from multispecies micro- (van den Brink et al., 1995) and mesocosm (Farmer et al., 1995) studies and in-situ bioassays (Schulz, 2003). Only a few field-relevant estuarine experiments have, however, been performed using microcosms (Chandler et al., 1997, deLorenzo et al., 1999).

Field studies have also been able to link changes in both freshwater community structure and pesticide exposure measured in the field (Schulz and Liess, 1999, Leonard et al., 2001, Bollmohr and Schulz, 2009). The adverse effects of toxicants on the community structure can be quantified through various means, including measuring the species richness (number of species) or the species diversity (equitability in species abundance among species). However, only recently has the Principal Response Curve (PRC) approach been shown to be valuable in statistically analyzing multiple endpoints (Den Besten & van den Brink 2005) and has been used successfully for biomonitoring in freshwaters by other authors (van den Brink and ter Braak, 1999; van den Brink *et al.*, 2009; Leonard *et al.*, 2001; Bollmohr and Schulz, 2009). Only a few biomonitoring studies have, however, established a link between pesticide exposure and effect in estuaries (Warwick *et al.*, 1990) and the PRC was never applied within these studies.

Most of the risk assessment studies focus on dissolved pesticides, due to their higher availability towards organisms. However, this is not justified in TOCE since many pesticides are accumulated over a long time (Chapman and Wang, 2001) and benthic organisms, including mainly deposit feeders are preferentially exposed to the particle-bound pesticides (Chandler *et al.*, 1997). Organisms living in the interface between water and sediment (mixture of epi-benthic, hyper-benthic and demersal meiofauna and zooplankton) are especially exposed to particle-bound contaminants (Chandler *et al.*, 1997) by the active ingestion of potentially toxic food particles. As a result, bioconcentration of insecticides in meiobenthos and other epi-benthic and hyper-benthic organisms cause significant ecotoxicological effects, even at relatively low aquatic concentrations (Chandler *et al.*, 1997).

For many juvenile fishes and shrimps, meiobenthos, and particularly harpacticoids copepods, are a significant food source (Coull, 1990). They also play important roles in marine ecological and physicochemical processes, such as cycling and remineralisation of organic carbon and the transfer of energy from primary producers to top trophic levels (Hicks and Coull, 1983; Coull, 1990).

Advantages of using meiobenthos in contaminant assessment studies include their high abundance and species diversity, short generation time (fast turnover rates), ubiquitous distribution, and moderate-to-high sensitivity to contaminants (Kennedy & Jacoby, 1999). Although they play a vital role in the ecological functioning of an estuary, they are still underrepresented in pollution studies (Coull & Chandler, 1992).

The effect assessment in the current study has been performed in order to understand the temporal (Chapter 4, 6) and spatial (Chapter 4) variability in benthic community structure and to determine which variable(s) are driving the dynamic of the community structure in an impacted estuary (Lourens River estuary) and in a natural estuary (Rooiels River estuary) (Chapter 6). The assessment was performed using Species

General Introduction

Sensitivity Distribution (Chapter 3), biomonitoring in the field with the use of multivariate statistics (Chapter 4, 6) and laboratory sediment toxicity tests under field-relevant conditions.

1.6 Towards a risk based protection of estuaries

The South African National Water Act (Act 36, 1998, implemented in 1999) makes provision for a Reserve to be determined prior to authorization of water use. The Reserve is "the quantity and quality of water required to satisfy basic human needs, considering both present and future needs and to protect aquatic ecosystems in order to secure ecological sustainable development and use of the resource". Within the Reserve Determination it is required to measure "toxic substances" in the sediment along the estuary in order to derive the status of the estuary and therefore the flow requirements of the system (DWAF, 2004). However, the measurement of toxins is often neglected. A recent literature review on the available information on TOCE's still does not mention any results on organic chemicals or heavy metals (Whitfield and Bate, 2007). The current study will clarify the need to measure particle-bound pesticides as a measure of deterioration and will identify the pesticides posing the highest risk in order to prioritise substances within TOCEs. Although it is required to measure "toxic substances", only limited water guideline values exist for freshwater systems (DWAF, 1996) and no guideline values exist for estuarine/ marine waters. No sediment quality guidelines exist either. Due to the lack of sediment toxicity data in South Africa, the only short term option would be to apply the EqP methodology (DiToro, 1991).

For the purposes of the preliminary determination of the Ecological Reserve in South Africa, the Reference Condition of an estuary refers to the ecological status that it would have had:
- when receiving 100% of the natural MAR
- · before any human development in the catchment or within the estuary
- · before any mouth manipulation practices (e.g. artificial breaching)

Typically, the Reference Condition in an estuary refers to its ecological status 50 to 100 years ago. However, only limited historical long term data exists on the abiotic and biotic conditions of South African TOCEs. Although the generic definition of the Reference Condition states that this condition should refer to the natural unimpacted ecological

General Introduction

state of the estuary, it is important to realise that in some estuaries, changes have occurred that may be irreversible (Whitfield and Bate, 2007). To determine the ecological reserve for an estuary the Reference Condition is compared with the present state by using an estuarine health index. The index entails water quality along with hydrology, hydrodynamics/ mouth condition/ physical habitat alteration, and biotic components (microalgae, macrophytes, invertebrates, fish and birds). However, exposure and effect assessment of particle-bound pesticides is hardly considered.

Many estuarine and coastal management initiatives worldwide, e.g. in North America, Europe and Australia, are required to derive and use environmental quality indices, e.g. implementation of the European Water Framework Directive (European Commission, 2000) and the UC Clean Water Act (USEPA, 2002). To be able to see changes we need to understand the basic natural conditions. There is a significant variability in seasonal and spatial aspects of estuaries, including the processes within the estuaries. Little is known about the behavior of different natural stressors (as salinity and temperature), anthropogenic stressors (as TSS, pesticides and nutrients) and the response of estuarine communities and thus variability of estuaries may be a significant source of error and need to be understood before conducting a risk assessment (Chapter 6).

1.8 Objectives and Structure of the Thesis

The research described in this thesis was designed to yield information on the impact of particle-bound pesticide on organisms living in the interface between sediment and water column in a temporarily open estuary. It was hypothesised that natural variables such as salinity and temperature and anthropogenic stressors such as particle-bound pesticides contribute to the variability of the system. This hypothesis was addressed through various studies, since many authors suggest a multiple line of evidence approach (Sediment Quality Triad) (Chapman, 1996) to assess the potential risk posed to benthic invertebrates by particle-bound contaminants. This approach is necessary due to the variability in sediment type, contaminant distribution and spatial and temporal variability within the ecosystem (Livingston, 1987), in particular within TOCEs. Accordingly, use of contamination exposure (Chapter 3), responses in the field (Chapter 4, 6) and assessment of laboratory exposures to field relevant concentrations (Chapter 5) remains the most accurate way to account for contaminant availability and effects.

General Introduction

Chapter 6 contributes to understand the temporal variability and driving variables in a "natural" compared to an "impacted" estuary, since no long-term data are available on the dynamics of natural variables, anthropogenic stressors and community changes. This understanding is very important in order to derive the correct Reserve for each estuary in South Africa.

Chapter 1: **General Introduction**

Chapter 2: **Study area**

Details are provided on the study area, general sampling methods used and chemical analysis of pesticide concentrations.

Chapter 3: **Probabilistic risk assessment and use of SSD**

The seasonal fluctuation of pesticide concentrations and their risk towards freshwater and marine communities in a TOCE was investigated within this study during a period of two years, based on
- the detection of particle-associated pesticides,
- calculation of dissolved pesticides using the EqPT,
- prediction of HC5 for marine and freshwater organisms and
- comparison of these values with threshold values suggested by international water quality guidelines (CCME, 1999; ANZECC, ARMACANZ, 2000; US EPA, 2002).

Chapter 4: **Identification of temporal and spatial variability of stressors and response indicators.**

The study examines the spatial and temporal variability in physico-chemical parameters, particle-bound pesticide concentrations and meiobenthic abundance during the dry season; and identifies the pesticides posing the highest risk towards the meiobenthos

General Introduction

community by comparing two runoff events, differing in their change in pesticide concentration and environmental variables.

Chapters 5: **Interaction between chlorpyrifos exposure, salinity change and salinity adaptation in a sediment toxicity test.**

The sediment toxicity test was performed with the identified pesticide posing the highest risk and the identified taxa, responding the most to pesticide exposure from the previous chapter.

The aim was to determine the interactive effect of pre-adaptation to varying salinities, salinity decrease, and CPF exposure on the survival rate of female and male *Mesochra parva*. We hypothesised that pre-adaptation to fluctuating salinities would lead to enhanced survival when exposed to a combination of CPF exposure and hypoosmotic stress.

Chapter 6: **Driving variables under Reference conditions compared to disturbed conditions**

The study examines the driving environmental variables (including natural and anthropogenic stressors) in a natural system versus an impacted system in order to identify how these variables change the community structure of the flocculent layer.

Chapter 7: **General Discussion**

CHAPTER 2

STUDY AREA

Study area

2.1 INTRODUCTION

This study was conducted in two estuaries along a 30km stretch of the False Bay coastline (Figure 2.1). False Bay is situated in the Western Cape with 30 km length and width, its volume is approximately 44.6 km^3 with a mean depth of 41m (Spargo, 1991).

Eleven river catchments drain into False Bay. All eleven catchments are small, with the result that river flow is very sensitive to rainfall, with peak flows occurring in winter. Only four of these estuaries have estuaries as it is described by Whitfield, 1992. The other rivers discharge through mouths displaying few estuarine characteristics. Agricultural impacts on estuarine ecosystems are very important especially in the Western Cape, since the estuarine catchment in this province showed the highest percentage of agriculture (37%) (Harrison *et al.*, 2001).

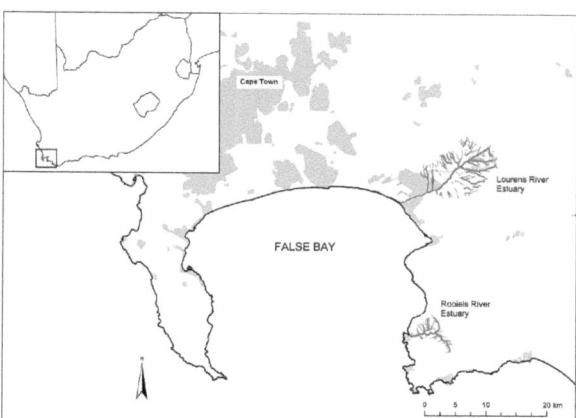

Figure 2.1 False Bay and the location of the two studied temporarily open estuaries.

The hydrodynamic of the bay is dominated by the bi-directional wind regime prevailing over the bay comprising predominantly SE winds in summer and NW winds in winter (Figure 2.2). Under S to SE wind conditions during summer both the surface and deep flows are generally clockwise. NW winds result in generally weaker flows with an anticlockwise tendency at the surface but a cyclonic tendency in the deeper waters.

Study area

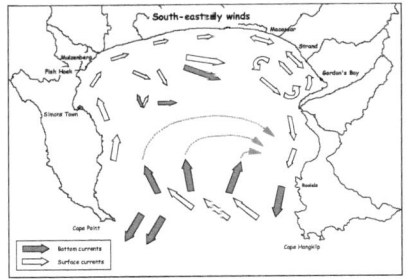 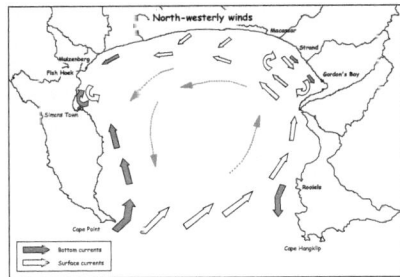

Figure 2.2 Direction of current during summer and winter within False Bay (Taljaard *et al.*, 2000)

2.2 STUDY ESTUARIES

2.2.1 Lourens River estuary

The Lourens River mouth is located at 34°06'S and 18°49'E in the northeast corner of False Bay. The river (20 km long) rises in the Hottentots Holland Mountains, flows through intensive agricultural areas and through the town Somerset West, after which it enters the False Bay at the town Strand (Cliff and Grindley, 1982). The estuary is about 0.710 km^2 in size with a tidal range of 1.48 m. Mean Annual Runoff is approximately 122 x 10^6 m^3 (Whitfield and Bate, 2007). Land-use in the Lourens River sub-catchment of 140 km^2 consists of forestry, agriculture, residential areas and light industries (Figure 2.3). A large section of the upper catchment is privately owned agricultural land with vineyards and apple, pear and plum orchards on which pesticide application takes place between August and mid February before fruit harvest.

Within the Lourens River estuary four sites were selected for various measurements. The location is shown in Figure 2.4, the exact coordinates and use in various chapters is shown in Table 2.1.

Study area

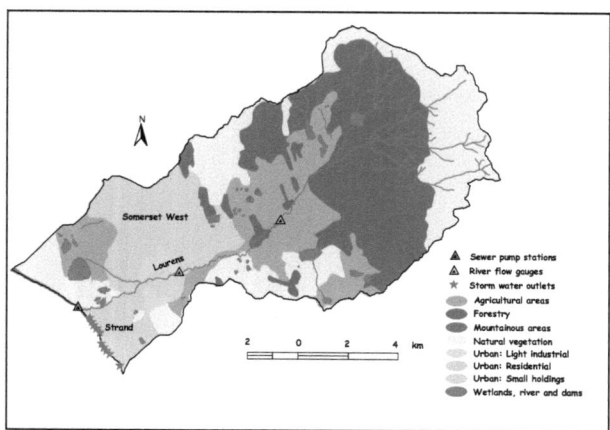

Figure 2.3 Lourens River catchment (Taljaard *et al.*, 2000)

Table 2.1 Site selection with its coordinates and the use in various chapters.

Study site	coordinates	chapter
Upper reaches	34°6'1.70"S 18°49'14.25"E	4
Middle reaches	34°6'3.48"S 18°48'56.48"E	4
Lower reaches	34°6'2.38"S 18°48'49.73"E	3,5,6
Outflow	34°5'58.44"S 18°48'37.34"E	3,6

Figure 2.4 Site selection within the Lourens River estuary (Google Earth, Map Data © 2008; © 2008 Europa Technologies)

2.2.2 Rooiels River estuary

The Rooiels River mouth is situated at 34°18'S and 18°49'E (1:50 000 Sheet 3418 BD). The Rooiels River catchment lies within the southern extension of the Hottentots Holland Mountain reserve (1:50000 Sheet SE 35/17 ½) and flows through the Kogelberg Nature Reserve (Figure 2.5). The total distance from the river mouth to the end of the longest tributary is only 9km. (Whitfield and Bate, 2007). The estuary is about 0.122 km² in size with a tidal range of 1.48 m. The Mean Annual Runoff is 10 x 10^6 m³. The entire Rooiels catchment (21 km²), lies within a nature reserve and has no associated agricultural catchments (Heydorn and Grindley, 1982), but small holiday residential areas (Figure 2.5).

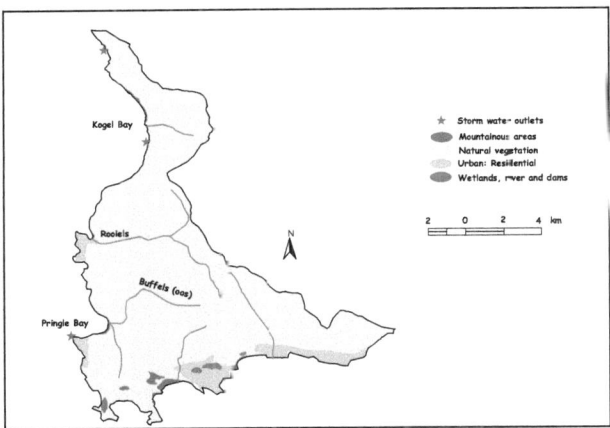

Figure 2.5 Rooiels River catchment (Taljaard et al., 2000)

Two sites were selected within in the Rooiels River estuary. The location is shown in Figure 2.6, the exact coordinates and use in various chapters is shown in Table 2.2.

Table 2.2 Site selection within the Rooiels River estuary with its coordinates and the use in various chapters.

Study site	coordinates	chapter
Upper reaches	-	-
Middle reaches	-	-
Lower reaches	34°17'58.74"S 18°49'22.61"E	6
Outflow	34°17'56.02"S 18°49'10.83"E	6

Study area

Figure 2.6 Site selection within the Rooiels River estuary (Google Earth, Map Data © 2008; © 2008 Europa Technologies)

CHAPTER 3

TEMPORAL VARIABILITY IN PARTICLE-ASSOCIATED PESTICIDE EXPOSURE IN A TEMPORARILY OPEN ESTUARY, WESTERN CAPE, SOUTH AFRICA

Published:

S. Bollmohr, J.A. Day, R. Schulz

(Chemosphere, Vol.68, pp. 479-488, 2007)

3.1 INTRODUCTION

The vast majority (71%) of South Africa's 258 estuaries are currently temporarily open/closed (TOCE), remaining closed to the ocean during the dry season while exhibiting an open phase with a duration proportional to the intensity of runoff during the rainy season (Whitfield, 1992). Suspended particles, originating from runoff in the upper catchment and transported into estuaries, can accumulate especially in TOCEs during low flow season, characterised by very little exchange with the sea. These particles, entering the Lourens River (Western Cape ;South Africa) have been shown to be associated with high concentrations of pesticides like chlorpyrifos (924 µg/kg), total endosulfan (12082 µg/kg) and prothiofos (980 µg/kg) (Schulz, 2001). Equilibrium partitioning theory (EqP) models (DiToro et al. 1991) and field studies (Domagalski & Kuivila, 1993) indicate that due to the high octanol-water partition coefficient of many common-use pesticides, most of the pesticides will deposit bound to the organic and clay fractions of sediments (Green et al., 1996). The degradation time of particle-associated pesticides is much greater than for aqueous phases, with some half-lives being greater than 100 d (Pait et al., 1992). Additionally, the residence time of pesticides associated to particles in estuaries is longer than pesticide dissolved in the water (Bergamaschi et al., 2001). The introduction of pesticides associated with particles may therefore increase exposure times to estuarine organisms.

Data on particle-associated pesticides are particularly relevant for estuaries; however, pesticides associated with suspended particles may have different environmental effects than dissolved pesticides. They might be particularly toxic to organisms living in the interface between sediment and water (flocculent layer) but are not necessarily bioavailable to organisms living in the water column. According to the equilibrium partitioning approach, the critical factor controlling sediment toxicity is the concentration of contaminant in the sediment pore water. Usually, no ecotoxicological data for sediment-dwelling marine organisms are available, and only a few values for freshwater sediment-dwellers can be found in the literature. Measuring concentrations of pesticides in the water can be a logistic problem due to the frequent change in the water level. The EqPT has been used in many studies to determine toxicity of particle bounded pesticides (Ankley et al., 1994; Green et al.,1996; Chandler and Green, 2001; Villa et al., 2003b). Furthermore, legislators are urged to use this approach to estimate maximum permissible concentrations from aquatic toxicity data, due to the lack of sediment toxicity data (European Commission, 2002; van Beelen et al., 2003)

Species can vary significantly in their sensitivity towards pesticides, and this variation can be described by constructing a species sensitivity distribution (SSD) (Posthuma et al., 2002). Species sensitivity distributions are used to calculate the concentration, at which a specified proportion of species within a community will be affected, referred to as hazardous concentration for 5% of species (HC5) (Forbes and Calow, 2002; Maltby et al., 2005). Within an estuary the salinity fluctuates daily and seasonally due to tidal influence and different freshwater flow conditions. Marine species tend to be more sensitive to pesticide compounds than freshwater species and the extrapolation of toxicity towards freshwater taxa to toxicity towards marine taxa is unclear. Insufficient toxicity data often exist for marine organisms to estimate a marine HC5. (Wheeler et al., 2002). Exposure and effect assessment of pesticides within estuaries is rare worldwide (Scott et al., 1999) and no information is available regarding pesticide concentrations and their toxicity in South African estuaries. Exposure assessment will provide an understanding of pesticide contamination and the ecotoxicological relevance within the whole catchment. The seasonal fluctuation of pesticide concentrations and their risk towards freshwater and marine communities in TOCEs was investigated within this study during a period of two years, based on the detection of particle-associated pesticides, calculation of dissolved pesticides using the EqP theory, prediction of HC_5 for marine and freshwater organisms and comparison of these values with threshold values suggested by international water quality guidelines (CCME 1999; ANZECC, ARMACANZ, 2000; US EFA, 2002).

3.2 MATERIAL AND METHODS

3.2.1 Study Area

The Lourens River estuary is a typical TOCE (Whitfield, 1992). It is located in the Western Cape, which is characterised by a high rainfall in winter (94.1 mm per month) and a dry summer (30.2 mm per month in 2001/2002 and 11.9 mm per month (2002/2003)) (Figure 3.1), as is characteristic of the regions Mediterranean climate. The estuary enters into False Bay, a nearly 40km wide bay southeastern of Cape Town.

The Lourens River mouth is situated at S34°06' and E18°49', the river length from the origin to the outflow is approximately 20 km. The whole estuary has a volume of approximately 0.710 km^3 and is characterised by a narrow outflow channel, a wide slow flowing middle reach and a narrow fast flowing upper reach. The study site was situated at the bottom of the middle reaches. A large section of the upper catchment is privately owned agricultural land with vineyards and apple, pear and plum orchards. The lower reaches of the total catchment area of 92 km³ are residential and light industrial.

3.2.2. Pesticide application and seasonal weather data

Within the investigation period (June 2001 to May 2003) the seasonal averages (spring, summer, autumn and winter) were determined (Figure 3.1) from the total monthly rainfall and the monthly average air temperature. Organophosphates such as chlorpyrifos (686 kg/ha) and prothiofos (87 kg/ha), organochlorines such as endosulfan (158 kg/ha) and pyrethroids such as cypermethrin, (8kg/ha) and fenvalerate (5 kg/ha) are frequently applied to pears, plums and apples between August and February before fruit harvest (Dabrowski et al., 2002) with a high frequency of application during the summer season from November to January (Schulz, 2001).

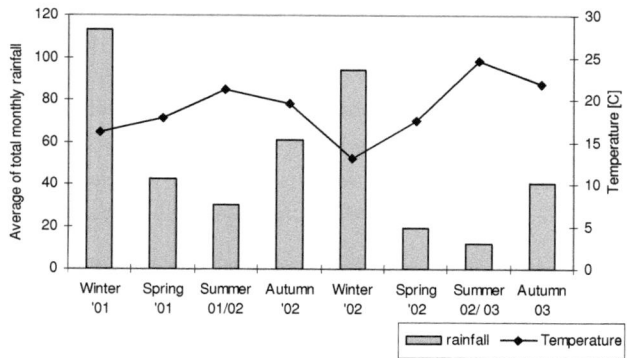

Figure 3.1 Seasonal rainfall and air temperature during the study period from 2002 to 2003 (data obtained from the Verelegen weather station at 80m altitude).

3.2.3. Physicochemical measurements

The flow was measured every meter across the outflow of the estuary using a flow meter (Hoentzsch Co., Waiblingen, Germany) and the discharge was calculated. Conductivity and temperature were measured at the surface and at the bottom to determine stratification trends within the estuary, using a conductivity meter LF 91 (WTW GmbH, Weilheim, Germany). The salinity was obtained by converting the conductivity values into salinity using an automatic converter (Fofonoff and Millard, 1983). The total organic carbon (%) was measured using a quantitative method which is based upon the indiscriminant removal of all organic matter followed by gravimetric determination of sample weight loss (ASTM, 2000). Four sediment samples (top 15cm) per site (within a 1m² transect) were taken and analyzed.

3.2.4. Pesticide Analysis

Samples of insecticides (one sample per sample event) associated with suspended particles were accumulated continuously by a suspended-particle sampler (Liess and Schulz, 1996), from which they were collected at 14-days intervals. The suspended-particle sampler consisted of a plastic container (500 ml) with a screw-on lid containing a hole (2 cm in diameter), which contained an open glass jar that was stored directly under the hole in the lid. The samplers were stored approximately 5 cm above the river bed by being attached to a metal stake which was fixed into the sediment.

Suspended particles samples were extracted twice with methanol and concentrated using C18 columns. Insecticides were eluted with hexane and dichlormethane and were analysed at the Forensic Chemistry Laboratory, Department of National Health, Cape Town. Measurements were made with gas chromatographs (Hewlett-Packard 5890, Avondale, PA, USA) fitted with standard Hewlett-Packard electron-capture, nitrogen phosphorus and flame-photometric detectors, with a quantification limit of 0.1µg/kg (Schulz et al., 2001) and overall mean recoveries were between 79% and 106% (Dabrowski et al., 2002). Further details of extraction and analysis of pesticides are described in Dabrowski et al. (2002).

3.2.5. Calculation of predicted pore-water concentration

Mean seasonal sediment concentrations for the detected pesticides were calculated. Predicted pore-water concentrations were calculated using the equilibrium partitioning equation (Di Toro et al., 1991) (Eq. (1)),

$$C_w = C_s / (f_{OC} K_{OC}) \tag{1}$$

where C_w is the estimated pore-water concentration (µg/L), C_s is the measured sediment concentration (µg/kg), f_{OC} is the fraction of organic carbon in the sediment, and K_{OC} is the organic carbon/water partitioning coefficient for the chemical of concern. For particle-associated and aqueous phase pesticide concentrations the 90th percentile concentration was calculated, which is a way of providing estimation of proportions of the data that should fall above and below a given value, often used in exposure estimations.

3.2.6. Derivation of predicted no-effect-concentrations using Species Sensitivity Distribution (BurrliOz)

Acute freshwater and marine median lethality data (LC50) for seven insecticides were extracted from the U.S. Environmental Protection Agency AQUIRE database (http://www.epa.gov/ecotox/). The data used in the analysis were derived from acute assays from publications after 1980 conducted over periods from 24 to 96 h. If data from multiple studies on the same species were available, the geometric mean toxicity values were used to represent the species in the distribution. Toxicity values reported as a value greater than a certain concentration were excluded from the dataset. Where sufficient toxicity values were available (at least six), data for arthropods and for fish were analyzed separately. An acute-to-chronic assessment factor of 0.1 was applied to the data. The Burr Type III (BT III) method used here, is the distribution that is required by the Environment Protection Authority and is implemented by the Department of Water Affairs and Forestry (DWAF) to derive threshold values for organic chemicals within the South African water quality guidelines (DWAF, 1996). The key advantage of this method is that it selects based on the maximum likelihood method the distribution that best fits the toxicity data from a family of distributions rather than trying to apply a single distribution. However, if another distribution could not be found that fitted the data better than the log-logistic distribution, this distribution is used, with model fit being evaluated using the Anderson-Darling goodness-of-fit test. The method of Aldenberg and Jaworska (2000) was used to calculate lower (95% confidence) and upper (5% confidence) limits.

3.2.7. Risk assessment

Calculated peak and average concentrations of pesticides in the water were compared with water quality criteria (US EPA, 2002; CCME, 1999; ANZECC, 2000; US EPA, 2002). Furthermore the toxicity-exposure-ratio (TER) was calculated by dividing the calculated HC5 for each pesticide by the concentration in the water. The chronic (using average exposure concentration) and acute (using peak exposure concentration) TER were compared amongst the pesticides to determine which pesticide present in the sediment may most likely cause toxicity. A risk is indicated when the acute TER value is below 100 and the chronic TER is below 10. Insufficient data for sediment quality criteria were available to asses the risk of the measured pesticides associated to suspended particles. Thus only a comparison with literature data was possible.

3.3 RESULTS

3.3.1. Physico-chemical parameters

During the summer seasons, characterised by low amount of rainfall, the lowest discharges of 0.21- 0.23 m³/s were measured (Table 3.1; Figure 3.1). At the same time the salinity and temperature increased due to higher evaporation and air temperature. The highest salinity (5.1 ± 5.7 ppt at the surface and 14.7 ± 12.3 ppt at the bottom) and the highest temperature (23.7 ± 3.2 °C at the surface and 23.5 ± 3.8°C at the bottom) were measured during summer season 2001-2002. The estuary developed stratification in salinity and temperature during dry season due to the increasing depth around the summer season, lack of freshwater input, intrusion of seawater during spring tide and higher evaporation. The total organic carbon of the suspended sediments showed its highest values during autumn 2003 with 3.5 ± 0.12% and lowest during autumn 2002 with 0.86 ± 0.22%.

Table 3.1 Mean (± standard error) seasonal fluctuation in physicochemical characteristics measured in the Lourens River within the period 2001-2003.

	Unit	Spring 2001 (n=3)	Summer 2001/2002 (n=3)	Autumn 2002 (n=3)	Winter 2002 (n=3)	Spring 2002 (n=6)	Summer 2002/2003 (n=6)	Autumn 2003 (n=6)
Q(outflow)	m³/s	nm	0.23 ± 0.00	0.41 ± 0.16	2.7 ± 1.4	0.99 ± 0.23	0.21 ± 0.12	0.44 ± 0.37
Salinity (surface)	ppm	0.34 ± 1.1	5.1 ± 5.7	0.83 ± 0.71	0.12 ± 0.01	0.21 ± 0.14	2.90 ± 2.23	5.2 ± 4.3
Salinity (bottom)	ppm	2.3 ± 7.2	14.7 ± 12.3	14.5 ± 14.4	0.13 ± 0.01	0.33 ± 0.34	6.4 ± 3.8	17.0± 13.2
T (surface)	°C	18.2 ± 3.8	23.7 ± 3.2	20.2 ± 5.7	13.6 ± 1.2	18.5 ± 3.4	23.7 ± 1.0	19.9 ± 1.1
T (bottom)	°C	18.2 ± 4.0	23.5 ± 3.8	19.6 ± 3.6	13.6 ± 1.2	18.4 ± 3.3	24.9 ± 1.6	20.7 ± 1.8
TOC	%	1.1 ± 0.2	1.5 ± 0.3	0.86 ± 0.22	1.0 ± 0.00	1.2 ± 0.09	1.1 ± 0.14	3.5 ± 0.12

3.3.2. Particle-associated pesticide concentrations

Cypermethrin, fenvalerate, endosulfan α, endosulfan β, endosulfan sulfate, p,p-DDE, chlorpyrifos and prothiofos were found in varying concentrations during almost all seasons (Table 3.2). Some of the pesticides showed seasonal patterns with higher concentrations around the summer seasons. The highest total average pesticide concentration was detected in spring 2002 with 76.1 ± 67.3 µg/kg, followed by the spring season in 2001 with 43.5 ± 43.8 µg/kg. During the winter season 2002, all pesticides showed the lowest concentration with a total average concentration of 4.21 ± 2.38 µg/kg. The peak concentrations of the different pesticides differed temporarily from each other. The pesticides p,p-DDE, fenvalerate and prothiofos showed the highest concentration during spring 2002; endosulfan β and chlorpyrifos showed the highest concentrations during spring 2001; endosulfan α and endosulfan sulfate during summer 2002/2003 and cypermethrin during autumn 2002. However, fenvalerate, cypermethrin and prothiofos were only included into the measurement scheme from autumn 2002 onwards. Fenvalerate showed the lowest 90th percentile concentration of 0.64 µg/kg, followed by cypermethrin with a concentration of 2.84 µg/kg. Within the different endosulfan isomers, endosulfan β showed the lowest 90th percentile concentration (5.28 µg/kg), followed by the breakdown product endosulfan sulfate with 7.08 µg/kg and endosulfan α with 7.55 µg/kg. Chlorpyrifos showed a 90th percentile concentration of 19.6 µg/kg, followed by p,p-DDE with 31 µg/kg. The organophosphate prothiofos showed the highest concentrations of all particle associated pesticides detected in the Lourens River estuary with a 90th percentile concentration of 34 µg/kg.

3.3.3. Aqueous phase pesticide concentrations

The highest total average pesticide concentration in water of 0.81 ± 0.72 µg/l was detected during spring 2002, followed by the concentration during spring 2001 with 0.38 ± 0.32 µg/l. The order of pesticides showing the highest and lowest 90th percentile and average concentration changed substantially. cypermethrin< fenvalerate< p,p-DDE< prothiofos< endosulfan< chlopryrifos. Generally, calculated water concentrations were rather low with 90th percentile concentrations of 0.45 µg/l for chlopryrifos, 0.16 µg/l for endosulfan, 0.15 µg/l for prothiofos, 0.05 µg/l for p,p-DDE, 0.01 µg/l for fenvalerate and 0.00 µg/l for cypermethrin. Organic carbon contents and different organic carbon/water coefficients of the pesticides (Table 3.3) resulted in seasonal patterns differing from the particle-associated pesticide concentrations. The high organic carbon content during autumn 2003 (Table 3.1) resulted in the lowest concentration of pesticides dissolved in the water (0.02 ± 0.03 µg/l).

Table 3.2 Mean (± standard error) particle-associated pesticide concentrations (µg/kg) in the Lourens River estuary per season and 90th percentile of all values within the overall sampling period 2001-2003.

	Particle-associated concentration (µg/kg)							Aqueous phase concentration (µg/l)	
	Spring 2001 (n=5)	Summer 2001/2002 (n=6)	Autumn 2002 (n=3)	Winter 2002 (n=2)	Spring 2002 (n=3)	Summer 2002/2003 (n=5)	Autumn 2003 (n=3)	90th percentile (n=27)	90th percentile (n=27)	Average concentration (n=27)
Chlorpyrifos	16.3 ± 12.2	9.82 ± 14.2	7.42 ± 7.83	2.60 ± 0.85	4.96 ± 4.06	2.20 ± 1.60	0.42 ± 0.51	19.56	0.446	0.085 ± 0.075
Prothiofos	nm	nm	8.43 ± 2.27	0.13 ± 0.18	38.8 ± 33.5	2.81 ± 1.67	0.64 ± 0.85	34.0*	0.145*	0.032 ± 0.051
Cypermethrin	nm	nm	2.70 ± 1.00	0.36 ± 0.05	1.92 ± 1.67	0.84 ± 0.71	0.33 ± 0.55	2.84*	0.001*	0.000 ± 0.000
Fenvalerate	nm	nm	0.29 ± 0.05	0.03 ± 0.02	0.79 ± 0.88	0.33 ± 0.30	0.06 ± 0.10	0.64*	0.013*	0.004 ± 0.004
Endosulfan β	3.76 ± 4.15	3.59 ± 3.65	1.96 ± 2.34	0.05 ± 0.07	2.29 ± 2.42	1.03 ± 1.81	0.26 ± 0.43	5.28		
Endosulfan α	3.00 ± 3.85	3.38 ± 3.20	2.53 ± 2.73	0.04 ± 0.06	3.09 ± 3.63	2.40 ± 4.68	0.35 ± 0.58	7.55		
Endosulfan sulfate	4.10 ± 4.89	5.73 ± 5.57	3.56 ± 3.30	0.34 ± 0.22	3.17 ± 2.76	0.38 ± 0.86	0.28 ± 0.40	7.08		
Total endosulfan	10.9 ± 12.2	12.7 ± 12.4	8.04 ± 8.36	0.43 ± 0.35	8.56 ± 8.82	3.81 ± 7.34	0.90 ± 1.40	18.6	0.158	0.033 ± 0.024
p,p-DDE	16.4 ± 18.7	10.5 ± 12.3	15.8 ± 16.1	0.66 ± 0.93	21.1 ± 18.4	0.00 ± 0.00	0.29 ± 0.50	31.0	0.054	0.010 ± 0.009
Total average concentration (µg kg^{-1})	43.5 ± 43.8	33.0 ± 38.9	42.7 ± 32.3	4.21 ± 2.38	76.1 ± 67.3	9.99 ± 11.6	2.64 ± 3.91			
Total average concentration (µg l^{-1})	0.38 ± 0.32	0.27 ± 0.34	0.32 ± 0.24	0.05 ± 0.02	0.81 ± 0.72	0.11 ± 0.10	0.02 ± 0.03			

nm: not measured
The right column presents the 90th percentile of the aqueous phase concentrations (µg l^{-1}) calculated based on the measured particle-associated values
* n=15

Table 3.3 Chemical properties (ARS, 1995; ATSDR, 2000; Pesticide Manual) of the pesticides detected in the Lourens River estuary

Insecticide	Solubility @25°C (mg l^{-1})	log K$_{OC}$	log K$_{OW}$	DT$_{50}$ (soil) aerobic (days)	DT$_{50}$ (soil) anaerobic (days)	DT$_{50}$ (water) (days)	Vapour pressure (mPa)	Henry Law constant (Pa m^3 mol^{-1})	Reference
Prothiofos	0.07	4.43^b	5.70	-	-	-	0.3		Pesticide Manual (1994)
Chlorpyrifos	1.18	3.70	5.27	31	-	35	2.43	0.743	ARS (1995)
Cypermethrin	0.00041	5.49^c	6.60	46	46	24	0.00019	0.0426	ARS (1995)
Fenvalerate	0.002	3.74	6.20	142	-	45	0.00223	0.042 (n=2)	ARS (1995)
p,p-DDE	0.07$^#$	4.82	6.96	-	-	3133	2.09	1.02	ARS (1995)
Endosulfan α	0.53	4.13^d	3.83^d	23	35	-	0.0013	1.03E-3	ATSDR (2000)
Endosulfan β	0.28	4.13^d	3.83^d	25	37	-	0.0013	1.94E-3	ATSDR (20000
Endosulfan Sulfate	0.22	4.13^d	3.83^d	-	-	-	0.0013	2.65E-3	ATSDR (2000)

^{a} logK$_{OC}$ and logK$_{OW}$ derived from Sabljik et al. (1995).
^{b} K$_{OC}$ derived from the correlation logK$_{OC}$=1.17 + 0.49 * logK$_{OW}$ (Sabljik et al., 1995)
^{c} logK$_{OC}$ derived from Laskowski (2002)
^{d} Only one logK$_{OC}$ only available for endosulfan (USDA ARS Pesticide Properties Database,1995, http://www.arsusda.gov/acsl/services/) (ATSDR Pesticide Properties Database, 2000. http://www.atsdr.cdc.gov).

3.3.4. Toxicity of pesticides based on SSD

The predicted hazardous concentrations (HC5) for both marine and freshwater distributions are summarised in Table 3.4. The distributions for each pesticide were based on a range of species numbers with 6-78 species for freshwater and 6-28 species for marine datasets.

The comparison of HC5 for freshwater and marine organisms showed, that marine organisms are more sensitive to cypermethrin, endosulfan, chlorpyrifos and fenvalerate, with lower HC5. Comparing the sensitivity of arthropod and fish taxa towards each pesticide showed a higher sensitivity of arthropods (marine and freshwater) towards all pesticides apart from chlorpyrifos. No toxicity was found for prothiofos, thus no HC5 was calculated.

Cypermethrin showed the highest toxicity with $HC5_{freshwater}$ of 0.0033 µg/l and $HC5_{marine}$ of 0.0015µg/l. However, the calculated hazardous concentrations were not exceeded by the concentrations during all seasons. Chlorpyrifos showed the second highest toxicity with $HC5_{freshwater}$ of 0.0045µg/l and $HC5_{marine}$ of 0.0017µg/l and was greatly exceeded by the concentrations measured and calculated in the estuary apart from autumn 2003. The 90th percentile concentration of fenvalerate (0.01 µg/l) exceeded only the $HC5_{marine}$ of 0.0036 µg/l and thus poses a risk only towards marine organisms Endosulfan α, β and sulfate were summarised due to no specification in toxicity data and the dissolved concentrations greatly exceeded the HC5 for freshwater (0.03 µg/l) and marine organisms (0.02 µg/l) during all seasons apart from winter 2002 and autumn 2003. The lowest toxicity was determined for o,p-DDE with a $HC5_{freshwater}$ of 0.2 µg/l, which was not exceeded by the calculated concentrations in the water. Due to the lack of marine toxicity data no $HC5_{marine}$ could be determined for p,p-DDE and even the calculated $HC5_{freshwater}$ needs to be looked at with caution due to the low number of available data (n=6).

Table 3.4. Estimation of mean (50% confidence) hazardous concentration which affects 5% of freshwater ($HC5_{freshwater}$) and marine organisms ($HC5_{marine}$) derived from species sensitivity distribution for arthropods and fish separately exposed to pesticides in single-species acute tests (US EPA database AQUIRE) by using the programme BurrliOz (* not sufficient data to perform analysis).

	$PNEC_{freshwater}$ (µg/l)				$PNEC_{marine}$ (µg/l)		
	Freshwater organisms	Freshwater Arthropods	Freshwater fish	Marine organisms	Marine arthropods	Marine fish
Cypermethrin	0.0033 (0.00032, 0.028)	0.0024 (0.003, 0.018)	0.05 (0.0026, 0.0539)	0.0015 (0.0001, 0.0033)	*	*
No of taxa	57	38	16	6	5	1
Endosulfan	0.03 (0.0036, 0.158)	0.02 (0.004, 0.0239)	0.04 (0.0068, 0.307)	0.02 (0.0013, 0.0306)	0.02 (0.0006, 0.075)	0.06 (0.0003, 0.4469)
No of taxa	78	23	48	28	16	7
Chlorpyrifos	0.0045 (0.0015, 0.0074)	0.01 (0.0022, 0.0124)	0.0002 (0.00008, 0.0089)	0.0017 (0.0004, 0.0046)	0.0047 (0.0003, 0.0077)	0.0012 (0.0002, 0.0068)
No of taxa	76	52	21	25	10	13
p,p-DDE	0.2	*	*	*	*	*
No of taxa	6	2	3	1	1	0
Fenvalerate	0.01 (0.0008, 0.0299)	0.0035 (0.0202, 0.0123)	0.03 (0.0053, 0.0602)	0.0036 (0.0002, 0.0056)	0.0014 (0.00003, 0.0032)	0.03 (0.002, 0.0341)
No of taxa	25	8	15	18	7	10

3.4 DISCUSSION

3.4.1. Pesticide exposure along the Lourens River

The pesticide samples in this study were collected in the lower reaches of the estuary approximately 5 km from the pesticide entry and are thus expected to be lower than in upper stream reaches closer to the pesticide entry. Rainfall-induced runoff resulted in an increase in the levels of the pesticides chlorpyrifos and endosulfan up to levels of 245 and 273 µg/kg in suspended particles in upper stretches of the Lourens River (Dabrowski et al., 2002). The lower concentration of chlorpyrifos detected in the estuary of 19.6 and 18.6 µg/kg suggesting degradation, uptake by macrophytes or dilution along the Lourens River. However no cypermethrin, and fenvalerate concentrations were found in the upper reaches (Schulz, 2001; Dabrowski et al., 2002), whereas the estuary showed concentrations as high as 2.34 and 0.64 µg/kg suggesting an accumulation downstream the river due to enrichment in fine particles. It is evident that the K_{OC} of each pesticide and the magnitude of application strongly influenced the amount of exposure concentration in the aqueous phase. Chlorpyrifos e.g. was applied with the highest amount (686 kg/ha) and was found in highest concentrations (0.09 ± 0.08 µg/l) whereas cypermethrin was applied with the smallest amount (2 kg/ha) and was found in lowest concentrations (0.00 ± 0.00 µg/kg).

3.4.2 Exposure and toxicity of particle associated pesticides

3.4.2.1. Chlorpyrifos

Chlorpyrifos concentrations as high as 245 µg/kg have been reported in Buzzards Bay, Massachusetts, USA (Pait et al., 1992). Bergamaschi et al. (2001) measured concentrations associated with suspended particles of 2.1 µg/kg in a permanently open estuary (San Francisco Bay), which is far below the average (6.25 ± 5.48 µg/kg) and 90th percentile (19.56 µg/kg) detected in this study. In a creek channel in California Hunt et al. (2003) measured concentrations up to 3.2 µg/l. Chlorpyrifos has been consistently detected in upper Chesapeake Bay waters at concentrations as great as 0.19 µg/l (Bailey et al. 2000), which is lower than the 90th percentile chlorpyrifos concentration calculated in this study (0.45 µg/l). A single species toxicity test with a marine copepod *Amphiascus tenuiremis* showed acute effects after 96 h at levels ranging from 21 to 33 µg/kg (Chandler et al., 1997), which are close to the detected 90th percentile of 19.6 µg/kg from the present study. Another study by Chandler and Green (2001) indicated a higher sensitivity of *Amphiascus tenuiremis* copepodit and naupliar showing a decrease in production during a chronic full life cycle exposure to lower concentrations (11-22 µg/kg) of sediment associated chlorpyrifos. However, a predicted sediment quality criterion (SQC) for a maximum safe concentration of chlorpyrifos based on the equilibrium partitioning theory was calculated to be 31.2 µg/kg sediment which is below the 90th percentile concentration found in this study. The HC5 derived in this study is lower than the threshold values derived by US EPA water quality guidelines (2002), but higher than the threshold values derived by CCME (1999) and ANZECC, ARMACANZ (2000) water quality guidelines (Table 3.5). The HC5 derived by Maltby et al. (2005) for arthropods (0.07 µg/l) is 7 times higher probably due to the fact that an acute chronic ratio factor of ten was applied in the present study or due to the different origin of the data. The 90th percentile concentration of chlorpyrifos in the water calculated in this study was up to 450 times higher than the recommended threshold values suggested by international water quality guidelines (CCME, 1999; ANZECC, ARMACANZ, 2000; US EPA, 2002). However the 90th percentile sediment concentration of 19.6 µg/kg was lower than the threshold value suggested by US EPA (1998) and ANZECC, ARMACANZ (2000).

The higher toxicity of chlorpyrifos determined for marine species cannot be explained by the taxa composition of the datasets, because insects (which are more sensitive than fish) are less represented in the marine dataset. This correlates with the finding by Hall and Anderson (1995) stating that the toxicity of organophosphate insecticides appeared to increase with salinity. Additionally the threshold values for freshwater communities suggested by the US EPA (2002), CCME (1999) and ANZECC, ARMACANZ (2000) (Table 3.5) were up to four times higher than the threshold values for marine communities, indicating a higher sensitivity of marine communities. Chlorpyrifos was one of the pesticides in the estuary posing the highest risk towards aquatic life with one of the lowest acute $TER_{freshwater}$ of 0.01 (TER_{marine} 0.004) and the lowest chronic $TER_{freshwater}$ of 0.05 (TER_{marine} 0.02) (Table 3.6).

3.4.2.2. Prothiofos
This study is one of the first records of prothiofos detected in the environment. No other comparable data were found amongst the literature. Prothiofos showed one of the highest concentrations associated to suspended particles with average concentrations of 10.2 ± 16.4 µg/kg and 90th percentile concentrations of 34.0 µg/kg. Due to the high $K_{OC,}$ the concentrations in the water were lower than the chlorpyrifos concentrations; however, this pesticide may still have the potential to pose a risk towards estuarine organisms. No sufficient toxicity data were found to determine potential risks to freshwater and marine organisms. Although this insecticide is

currently not in use in Europe and United States, further work needs to be done in order to facilitate a risk assessment approach for countries like South Africa and Australia. According to www.pesticideinfo.org prothiofos is only registered in South Africa, New Zealand and Australia.

Table 3.5. Water and sediment quality criteria set by various international guidelines in comparison to HC5 concentration derived in this study for freshwater and marine organisms.

	Freshwater		Marine	
	Chlorpyrifos	Endosulfan	Chlorpyrifos	Endosulfan
HC5 values derived in this study	0.0045	0.03	0.0017	0.02
Water quality criteria (µg/L)				
US EPA Quality criteria for water (2002)	0.041	0.22	0.011	0.034
Canadian Environmental Quality guidelines (CCME, 1999)	0.0035	0.02	0.002	-
Australian and New Zealand guidelines for fresh and marine water quality ANZECC, ARMACANZ, 2000	0.001	0.01	0.001	0.01
			Chlorpyrifos	Endosulfan
Sediment quality criteria (µg/kg) in freshwater systems only				
US EPA Quality criteria for water (2002)			32.2	5.4
Australian and New Zealand guidelines for fresh and marine water quality ANZECC, ARMACANZ, 2000			53	0.3

3.4.2.3. Cypermethrin

No comparable concentrations of cypermethrin in estuaries were found in the literature. Cypermethrin and fenvalerate are pyrethroids and are therefore expected to be more toxic to arthropods and crustacean than to fish (Solomon et al., 2001), which is in agreement with the findings in this study. Clark et al. (1989) measured mortality rates from 28% (grass shrimp) to 85% (mysid) after a 4 day static exposure of 100µg/kg cypermethrin. The 10-day median LC50s of cypermethrin were 3.6 -23 µg/kg for *Hyaella azteca* and 13- 62µg/kg for *C. tentans* in 1% 3%, 13% OC sediments, respectively, due to its low bioavailability and binding potential of up to 99% (Maund et al., 2002). The 90th percentile concentration of 2.84 µg/kg was similar to the LC50 for the sediment with the lowest organic content for *Hyalella azteca*. Therefore concentrations measured in the sediment may pose a risk to organisms living in the interstitial.

Brock et al. (2000) suggested a safe concentration for surface water by calculating a $LOEC_{eco}$ of ≤ 0.07 µg/l, based on mesocosm studies. Giddings et al. (2001) conducted mesocosm studies and determined lowest-observed-adverse-effect concentrations (LOAEC) ranging from 0.03 µg/l (amphipods, isopods) to >1 µg/l (odonates, fish, snails). The HC5 (0.0015- 0.0033 µg/l) calculated in this study is lower than the suggested threshold values. However Maltby et al. (2005) reported a HC5 for freshwater arthropods of 0.003 µg/l which is very similar to the HC5 determined in this study (0.002 µg/l). No threshold values for cypermethrin are given by international guidelines. Due to the very high K_{OC} only very low concentrations were calculated for the aqueous phase. The high acute $TER_{freshwater}$ of 0.17 (TER_{marine} 0.03 marine) and chronic $TER_{freshwater}$ of 1.1 (TER_{marine} 0.03) suggesting that this pesticide posed a low acute and chronic risk towards aquatic life in the Lourens River estuary.

Table 3.6. Calculated chronic and acute toxicity exposure ratio (TER) for the pesticides detected in the Lourens River estuary using the calculated concentrations in the water and the calculated predicted no-effect-concentrations.

	TER_{acute}		$TER_{chronic}$	
	Freshwater	Marine	Freshwater	Marine
Chlorpyrifos	0.01	0.004	0.05	0.02
Cypermethrin	3.3	6.7	11	5
Fenvalerate	0.8	0.3	2.5	0.9
Endosulfan	0.2	0.1	0.9	0.6
p,p-DDE	3.7	-	20	-

3.4.2.4. Fenvalerate

Reported peak concentrations of fenvalerate associated with particles are 100 µg/kg (Chandler et al., 1994) in an estuary. Reported peak concentrations of fenvalerate in streams and estuaries are 0.1 to 6 µg/L (Schulz and Liess, 2001) for the aqueous phase, 100 µg/kg for fenvalerate associated with sediments (Chandler et al., 1994), and 302 µg/kg for fenvalerate associated with suspended particles(Schulz and Liess, 2001). All concentrations were much higher than the concentrations found in this study. Clark et al. (1989) suggested that fenvalerate concentrations of greater 0.1 µg/l would cause significant field mortality for the estuarine mysid *Mysidopsis bahia*. Aqueous-phase fenvalerate caused sublethal response of a trichoptera species at 0.0001µg/l (2 µg/kg) and lethal effects at 0.01 µg/l (2000 µg/kg) in a study by Schulz and Liess (2001), which correlates with the determined HC5 for freshwater arthropods in this study of 0.0035 µg/l, whereas the HC5 for arthropods derived by Matlby et al. (2005) of 0.013 µg/l is higher. The 90th percentile fenvalerate concentration of 0.01 µg/l was higher than the HC5 for freshwater and marine organism derived in this study and therefore fenvalerate was likely to pose a risk to the organisms.

3.4.2.5. Endosulfan

The active ingredients of technical endosulfan used for commercial product formulation consist of a racemic mixture of the isomers α-endosulfan and β-endosulfan in a ratio of between 2:1 to 7:3 (Wan et al., 2005). Thus in soils and water, the half-life for degradation time of α-endosulfan is about one to three months whereas that of β-endosulfan and endosulfan sulfate can be from two to six years depending on environmental conditions (Wan et al., 2005), which explains the higher 90th percentile of particle associated endosulfan sulfate detected in the Lourens River estuary. Endosulfan concentrations in the water of the Minjiang River estuary in Southeastern China were measured and resulted in concentrations of up to 0.066 µg/l for α-endosulfan, 0.215 µg/l for β-endosulfan and 0.135 µg/l for endosulfan sulfate (Zhang et al., 2003). Doong et al. (2002) measured endosulfan concentrations in the sediment of the Wu-Shi River estuary in Taiwan with up to 5.67 µg/kg α-endosulfan, 1.82 µg/kg β-endosulfan and 10.5 mg/kg endosulfan sulfate. All reported concentrations were exceeded by the 90th percentile derived from this study.

Wan et al.,(2005) determined a LC50 for endosulfan (α+β) for Daphnia (840µg/l), a salmonid fish (0.7 µg/l) and an amphipod (5.7 µg/l), suggesting fish being the most sensitive taxa. Leonard et al. (2001) determined a 10-d no observed effect concentration of 42 µg/kg for sediment-associated endosulfan for the epibenthic mayfly *Jappa kutera*. No effects on benthic macroinvertebrates were detected at either the population or the community level in response to peak endosulfan concentrations of 6.14 µg/l in the interstitial water. Another study by Hose et al. (2003) demonstrated toxicity of endosulfan concentrations in the water from 6.3 µg/l towards the mayfly *Atalopnlebia spp* (LC50 12.3–13.6 µg/l, LC10 6.3-8.2 µg/l). Scott et al. (1999) determined a LC50 of 1.01 µg/l for the estuarine grass shrimp *P. pugio*, which is exceeded by the 90th percentile concentration in this study. Chandler and Scott (1991) suggested that polychaete (*Streblospio benedicti*) populations in the field may be strongly depressed by sediment endosulfan concentration of 50µg/kg. However survival of the benthic copepod *Nannopus spec*. was only affected by 20% at concentration as high as 200µg/kg. Among the different isomers of endosulfan, α endosulfan is the most toxic of all endosulfan materials tested to non-target aquatic organisms (Wan et al., 2005). It is about 1.3, 6.6 and 58 times more toxic than β-endosulfan to *Daphnia*, salmonid fish, and amphipods, respectively. The study by Wan et al. (2005) also confirms that endosulfan sulfate, the transformation product of both isomers is as toxic as the parent compound to salmonid fish and crustaceans.

The above mentioned effect concentrations are much higher than the concentrations found in the sediment and calculated for the water within this study. However the threshold values suggested by CCME (1999) and ANZECC, ARMACANZ (2000) of 0.02 and 0.01 µg/l (Table 3.5), respectively are lower than the 90th percentile concentration of 0.16 µg/l indicating a risk within the Lourens River estuary. The 90th percentile concentrations of endosulfan measured in suspended particles exceeded the threshold value suggested by US EPA (2002) (5.4 µg/kg) and ANZECC, ARMACANZ (2000) (0.3 µg/kg). The higher toxicity of endosulfan determined for marine species is also reflected within the guidelines (ANZECC, ARMACANZ, 2000; US EPA, 2002). Endsulfan (together with chlorpyrifos) poses the highest risk towards aquatic organisms with a low acute $TER_{freshwater}$ of 0.1 (TER_{marine} 0.1) and a low chronic $TER_{freshwater}$ of 0.9 (TER_{marine} 0.6) (Table 3.6).

3.4.2.6. p,p-DDE

Lotufo et al. (2000) determined a 4-d LC50 of 10.99µg/l and a 10-d LC50 of 3.88µg/l for the amphipod Hyaella azteca. Another amphipod, Dipoeria spp, did not show a significant decrease in mortality after 28 days exposure to p,p-DDE concentration ranging from 2.29-20.2 µg/l. This study showed low concentrations in the water (0.01 µg/l), which did not exceed the HC5 for freshwater (0.2 µg/l). However the CCME guidelines (1999) suggest a threshold value of 1.42 µg/kg for freshwater systems and 2.07 µg/kg for marine systems which is up to 6 times lower than the average concentration p,p-DDE found in this study. According to the high acute and chronic TER calculated in this study (Table 3.6), p,p-DDE does not pose a risk to aquatic life. Even though the toxicity of p,p-DDE is far lower than DDT and other metabolites (Lotufo et al., 2000), it is highly persistent in natural systems, thereby posing a threat to biota (Gillis et al., 1995) due to the high potential of bioaccumulation (Landrum et al., 2005).

3.5. CONCLUSION

Most of the particle associated pesticide concentrations, detected in this study were lower than the concentrations found in the same river closer to the source, indicating degradation along the river system that requires further investigation. A higher risk during the spring season was identified with pesticide concentrations exceeding threshold values set by international guidelines, due to the higher application rate of pesticides and lower flow resulting in an accumulation of pesticides. During this season the salinity values increased, resulting in a higher abundance of more sensitive marine organisms. Chlorpyrifos and endosulfan showed the lowest TER and therefore posed the highest risk towards freshwater and marine communities in the Lourens River estuary. The difference between the toxicity of pesticides towards freshwater and marine organisms was consistent and the factor ranged between 1.5 and 2.8. However, this conclusion needs to be qualified for each pesticide and for the mixture of pesticides in view of the small dataset for some pesticides (e.g. fenvalerate or p,p-DDE). This will require marine data generation and potentially further development of test methods with marine or estuarine species.

The benchmarks (HC5s) derived in this study should not be used as the sole measure of sediment toxicity. Field studies and toxicity tests should be primary indicators of toxicity of sediments and require further attention. Benchmarks may be used to determine which chemical present in the sediment may most likely cause toxicity. This integrative approach allows a more accurate evaluation of adverse ecological impact, which is necessary in a baseline risk assessment. Given the uncertainty in the derivation of sediment guidelines, the approach is possibly justified for the provision of interim guideline values.

CHAPTER 4

SPATIAL AND TEMPORAL VARIABILITY IN PARTICLE ASSOCIATED PESTICIDE EXPOSURE AND THEIR EFFECTS ON BENTHIC COMMUNITY STRUCTURE ALONG A TEMPORARILY OPEN ESTUARY

Published

S. Bollmohr, P.J. van den Brink, P.W. Wade, J.A. Day, R. Schulz
(Estuarine, Coastal and Shelf Science (in press))

4.1 INTRODUCTION

Temporarily open/closed estuaries (TOCE) are periodically closed off to the sea due to the lack of freshwater input. They show significant natural spatial and temporal variability in physical and chemical conditions (such as flow, salinity, temperature or total suspended solids). This variability is likely to be reflected in changes in community structures (Kibirige & Perissinotto, 2003). Thus, trying to isolate fingerprints of contaminant effects is particularly difficult when their distribution is confounded with physical and chemical gradients. Methods of community description used to determine which variables and which taxa contribute to spatial and temporal differences between the sites chosen include: i. community indices (e.g. species richness and Shannon Diversity Index) and ii. a multivariate approach, namely Principal Response Curve. The Principal Response Curve (PRC) approach has been shown to be valuable in statistically analyzing multiple endpoints (Den Besten & van den Brink, 2005). The statistical methodology has been used already for biomonitoring by other authors (van den Brink and ter Braak, 1999; van den Brink et al., 2009).

The study area in this investigation is the Lourens River estuary (Western Cape, South Africa). Previous studies (Bollmohr and Schulz, 2009; Schulz et al., 2001) has been shown that pesticides are contributing most to the pollution and associated effects in the Lourens River. Particles entering the estuary were shown to be associated with high concentrations of chlorpyrifos (19.6 µg/kg), endosulfan (18.6 µg/kg) and prothiofos (34.0 µg/kg) (Bollmohr et al., 2007), with highest concentrations during the spring and summer season. The accumulation of these particles is enhanced by a number of factors and may show a spatial gradient in concentrations and effects during this season. These factors include change from fresh to saltwater, rise and fall of water level with the tides, decrease in flow within the middle reaches of the estuary, and influx of seawater.

This study focused on community changes of organisms living in the flocculent layer between sediment and water column, including epi-benthic, hyper-benthic and demersal meiobenthos and zooplankton organisms. Advantages of using meiobenthos in contaminant assessment studies include their high abundance and species diversity, short generation time, fast turnover rates, ubiquitous distribution, and moderate-to-high sensitivity to contaminants (Kennedy & Jacoby, 1999). More than 98% of all meiobenthic copepods live in the upper 1 to 2 cm oxic zone of muddy sediments (flocculent layer) and thus may be particularly affected by exposure to particle-bound pesticides (Chandler et al., 1997). Deposit- and suspension-feeding infauna living in contaminated sediments actively ingest potentially toxic food particles and circulate contaminated water. As a result, bioconcentration of insecticides in meiobenthos and other epi-benthic and hyper-benthic organisms cause significant ecotoxicological effects, even at relatively low aquatic concentrations (Chandler et al., 1997). Copepods and meiobenthos as a whole are underrepresented in pollution studies (Coull & Chandler, 1992), yet they play a major role in estuarine carbon cycling and food webs. Their variability (daily and seasonal) can, however, be very high and difficult to correlate with environmental factors because of multifactorial relationships.

Many studies of meiobenthic community structure focus on the natural distribution and its driving variables like sediment composition (Sherman & Coull, 1980), salinity (Miliou, 1993) and food source (e.g. Decho & Castenholz, 1986). To determine the effect of pesticides on meiobenthos abundances, various studies were performed using sediment toxicity tests (e.g. Chandler & Green, 2001; Bollmohr et al., 2009b) or microcosms (Chandler et al., 1997), but only a few studies using biomonitoring in the field (Warwick et al., 1990) have been performed. Some studies have focussed on seasonal fluctuations of meiofauna and zooplankton in South African estuaries in relation to the distribution of physical factors (Dye, 1983; Nozais et al., 2005), state of the estuary mouth (Kibirige & Perissinotto, 2003) and food web interactions (Perissinotto et al., 2000), but the current study is the first to include investigation of the effects of pollutants.

The objectives of this study are:

- to examine the spatial and temporal variability in physico-chemical parameters, particle-bound pesticide concentrations and meiobenthic abundance during the dry season; and
- to study the effects of particle-bound pesticides on the dynamics of the meiobenthos community by comparing two runoff events, differing in their change in pesticide concentration and environmental variables.

4.2 MATERIALS AND METHODS

4.2.1 Study area
The Lourens River estuary is a typical temporarily open-closed estuary (Whitfield, 1992). It is located in the Western Cape of South Africa, which has a Mediterranean climate and is characterised by high rainfall in winter and a dry summer. The estuary enters into False Bay, an approximately 40 km wide bay south-east of Cape Town. The Lourens River mouth is situated at S34°06' and E18°49' and the total river length is approximately 20 km. The entire estuary has a volume of approximately 0.710 km^3 and is characterised by a narrow outflow channel, a wide slow flowing middle reach and a narrow fast flowing upper reach. The study sites were situated in the middle reaches (lower site) and the upper reaches (upper site) of the estuary. A large section of the upper catchment comprises privately-owned agricultural land with vineyards and apple, pear and plum orchards. The lower reaches of the total catchment area of 92 km^2 consist of residential and light industrial.

4.2.2 Pesticide application and seasonal weather data
The four-month investigation period (28 November 2002 to 28 March 2003) lay within the dry summer season with infrequent rainfalls (Figure 4.1). The second, much heavier rainfall event on the 23rd March (30.2 mm) took place in the low-spraying season.

The following pesticides (with corresponding loads) are applied to pears, plums and apples between August and February before fruit harvest (Dabrowski et al., 2002): organophosphates such as chlorpyrifos (686 kg/ha) and prothiofos (87 kg/ha); organochlorines such as endosulfan (158 kg/ha); pyrethroids such as cypermethrin (8 kg/ha) and fenvalerate (5 kg/ha). There is a particularly high frequency of pesticide application during the summer season from November to January – the "high-spraying season" (Schulz, 2001). Only rainfall events above 10mm are assumed to contribute to runoff with possible input of particle-bound pesticides (Schulz, 2001). Only two such events were recorded (Figure 4.1). The first rainfall event on the 15th January (11 mm) occurred during the high-spraying season.

Spatial and temporal variability

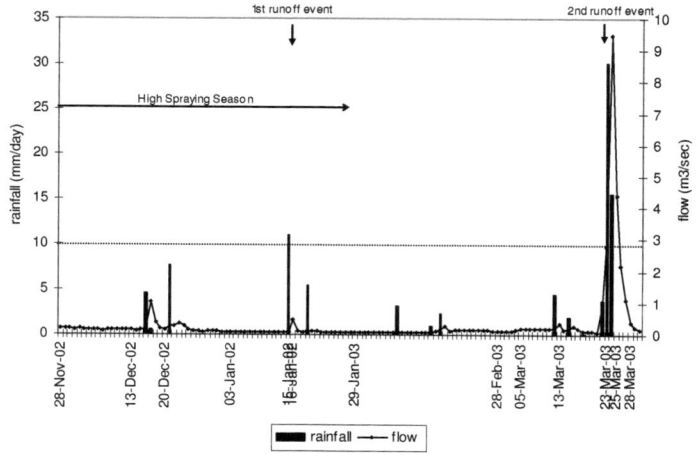

Figure 4.1 Daily rainfall (obtained from Verlegen weather station at 80m altitude) and flow data (obtained from DWAF gauging station G2H029A01) during the study period from November 2002 to March 2003. (The dotted line indicates the threshold of 10mm runoff above which runoff is assumed to take place; arrows indicating two runoff events above 10mm)

4.2.3 Physicochemical measurements

Various physicochemical parameters were measured during each sampling event. The flow was measured every meter across the cross-section of the estuary at the different sites using a flow meter (Hoentzsch Co, Waiblingen, Germany). Dissolved oxygen, pH, temperature and conductivity were measured with electronic meters from Wissenschaftliche Technische Werkstaetten GmbH, Weilheim, Germany. The salinity measurement was obtained by converting the conductivity values into salinity using an automatic converter (Fofonoff & Millard, 1983). Total Suspended Solids (TSS) was measured with a turbidity meter (Dr. Lange, Duesseldorf, Germany) and nutrients were detected by photometric test kits from Macherey & Nagel, Dueren, Germany. The total organic carbon (TOC) was measured using a quantitative method which is based upon the indiscriminate removal of all organic matter followed by gravimetric determination of sample weight loss (ASTM, 2000). Four sediment samples (top 15 cm) per site (within a 1 m^2 transect) were collected and analysed.

4.2.4 Pesticide analysis

Samples of insecticides (one sample per sample event) associated with suspended particles were accumulated continuously over 14-day periods by a suspended-particle sampler (Liess et al., 1996), from which they were sampled for analysis. The suspended-particle sampler consisted of a plastic container (500 ml) with a screw-on lid. A hole (2 cm in diameter) was cut into the lid. An open glass jar that was stored directly under the hole in the lid. The samplers were stored approximately 5 cm above the river bed by being attached to a metal stake which was fixed into the sediment. Samples of suspended particles were extracted twice with methanol, and the methanol-soluble fraction was concentrated using C18 columns. Insecticides were eluted with hexane and dichloromethane and were analysed at the Forensic Chemistry Laboratory, Department of National Health, Cape Town. Measurements were performed using gas chromatographs (Hewlett-Packard 5890, Avondale, PA, USA) fitted with standard Hewlett-Packard electron-capture, nitrogen phosphorus and flame-photometric detectors, with a quantification limit of 0.1 µg/kg and overall mean recoveries between 79 and 106% (Schulz et al., 2001). Further details of extraction and analysis of pesticides are described in Schulz et al. (2001).

4.2.5 Meiobenthic sampling and community analysis

A different sampling approach from sediment core sampling was applied to reduce the sampling error caused by small-scale variability in sediment (Livingston, 1987) and to sample the organisms living in the flocculent layer between sediment and water column, including epi-benthic, hyper-benthic and demersal organisms. A stretch of 10 m flocculent layer (width: 30 cm) was sampled actively just above the bottom sediment using a zooplankton net of 63 µm mesh size. Four replicates were taken and within each replicate the zooplankton and meiobenthos were identified to the lowest taxonomic level possible. Zooplankton and meiobenthos abundance was determined based on eight sub-samples. Sub-samples were derived using a matrix box with 64 chambers, of which 8 chambers were randomly selected for analysis. To assess temporal and spatial patterns of benthic faunal assemblages the following community metrics were calculated: density (Ind/m^3), no of taxa, H' diversity (Shannon & Wiener, 1948).

Spatial and temporal variability

4.2.6 Statistics

The differences in community composition between the data sets from the upper and middle reaches were visualised by PRC (Principal Response Curves; Van den Brink & Ter Braak, 1999) using the CANOCO software package version 4.5 (Ter Braak & Smilauer, 2002). PRC is based on the Redundancy Analysis ordination technique (RDA), the constrained form of Principal Component Analysis. The analysis produces a diagram showing time on the x-axis and the first Principal Component of the differences in community structure on the y-axis. This yields a plot showing the most dominant deviations with respect to time between the two sites, with the midstream site as reference. The species weights are shown in a separate diagram, and indicate the affinity the species have with respect to this difference pattern (Figures 4.2 – 4.4). Species with a high positive weight are indicated to show differences similar to the pattern indicated by PRC, and those with a negative weight show differences opposite to the pattern indicated by PRC. Species with near-zero weights show either a pattern very dissimilar to the differences indicated by PRC or indicate that there are no differences at all.

The data sets of the physico-chemical variables and pesticides were also analysed using PRC. In these analyses the variables were centred and standardised before analysis to account for differences in measurement scale (Kersting & Van den Brink, 1997). Further applications in PRC for analysis of monitoring data are described in greater detail in Den Besten & van den Brink (2005).

4.3 RESULTS

4.3.1) Temporal and spatial variability between sites

The PRC diagram indicates differences in environmental variables, pesticide concentrations and community structure between the two sites within the estuary, explaining 37%, 35% and 31%, respectively, of the total variation (Figures 4.2, 4.3 and 4.4). However, 63%, 65% and 69% of the total variation could be attributed to differences in environmental parameters, pesticides and community structure between sampling dates. This indicates a higher temporal variability than spatial variability in the system. The variability is mostly explained by TOC, salinity, and the pesticides total endosulfan ($\alpha+\beta$, sulfate) (Total END) and chlorpyrifos (CPF). Taxa contributing most to the variability are *Mesochra* and the freshwater species *Dunhevedia* and *Thermocyclops*.

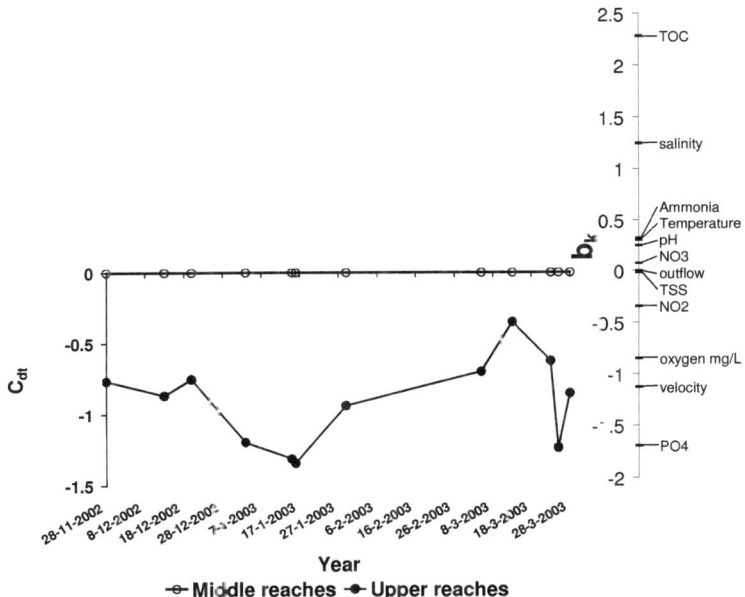

Figure 4.2 Diagram showing the first component of the PRC of the differences in measured physico-chemical parameters between the middle and upper reaches. Sixty-three percent of the total variation in physico-chemical parameter composition could be attributed to differences between sampling dates, the other 37% to differences in physico-chemical parameter composition between the upstream and downstream sampling site; 47% of the latter is displayed in the diagram. The parameter weights shown in the right part of the diagram represent the affinity of each parameter with the response shown in the diagram.

Spatial and temporal variability

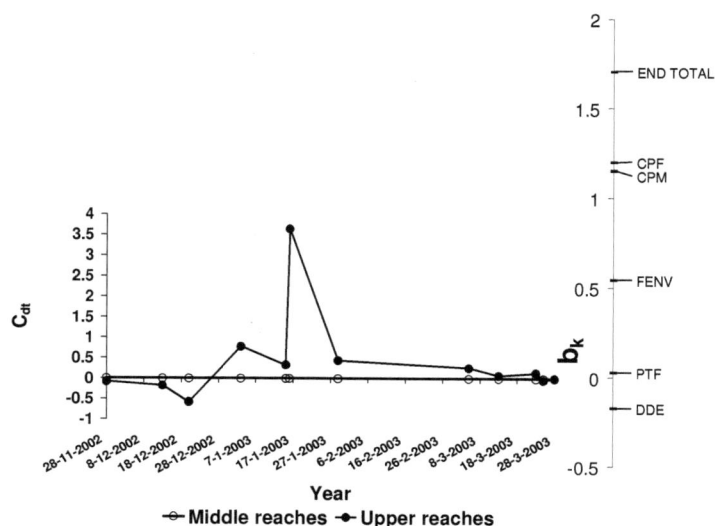

Figure 4.3 Diagram showing the first component of the PRC of the differences in measured pesticide concentrations between the middle and upper reaches. Sixty-five percent of the total variation in pesticide composition could be attributed to differences between sampling dates, the other 35% to differences in pesticide composition between the upstream and downstream sampling site; 65% of the latter is displayed in the diagram. The pesticide weights shown in the right part of the diagram represent the affinity of each pesticide with the response shown in the diagram.

4.3.2) Environmental variables and particle-bound pesticides

4.3.2.1) General spatial variability

Only the environmental variables which are significantly different between the two sites are shown in Table 4.1. Variables like pH, TSS, grain size, nitrate, nitrite and ammonia did not differ significantly between sites. The average surface salinities of 2.16 ± 2.60 ppt and bottom salinity of 18.3 ± 15.1 ppt within the middle reaches of the estuary were significantly higher than the average salinities measured at the upper reaches (surface 0.13 ± 0.13 ppt; bottom 2.18 ± 3.54 ppt; p = 0.02 and 0.001, respectively).

Spatial and temporal variability

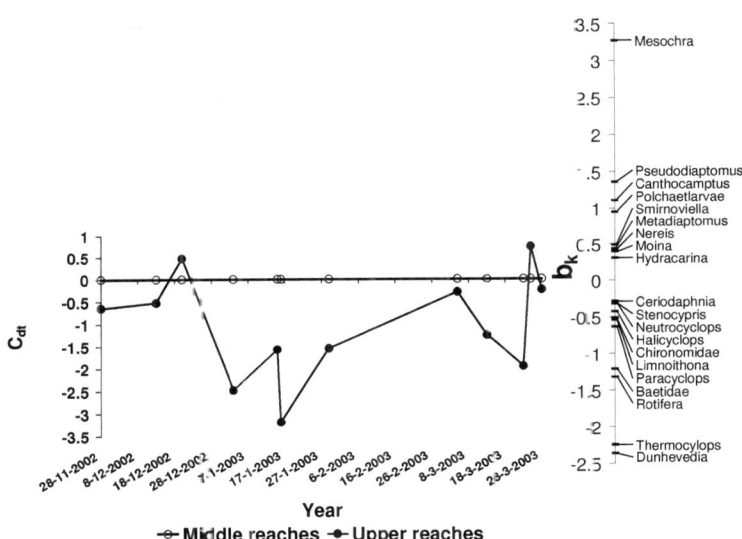

Figure 4.4 Diagram showing the first component of the PRC of the differences in species composition between the middle and upper reaches. Sixty-nine percent of the total variation in species composition could be attributed to differences between sampling dates, the other 31% to differences in species composition between the upstream and downstream sampling sites; 36% of the latter is displayed in the diagram. The species weights shown in the right part of the diagram represent the affinity of each species with the response shown in the diagram. For the sake of clarity, only species with a species weight larger than 0.25 or smaller than -0.25 are shown.

Spatial and temporal variability

Table 4.1 Physicochemical parameters measured within the upper and middle reaches of the Lourens River estuary (Only the variables which are spatially significantly different (p<0.05) are shown) during the study period November 2002 to March 2003.

		28.11.02	13.12.02	20.12.02	03.01.03	15.01.03	16.01.03	29.01.03	05.03.03	13.03.03	23.03.03	25.03.03	28.03.03	av ± STD
Salinity$_{surface}$ (ppt)	upper	0.12	0.11	0.11	0.19	0.12	0.10	0.28	0.45	0.01	0.01	0.01	0.01	0.13 ± 0.13
	middle	0.25	0.44	0.26	8.29	2.05	0.17	2.10	3.25	3.65	0.01	0.12	5.37	2.16 ± 2.60
Salinity$_{bottom}$ (ppt)	upper	0.12	0.11	0.11	10.1	0.12	0.10	3.43	3.87	8.17	0.1	0.1	0.01	2.18 ± 3.54
	middle	6.12	9.21	0.26	42.7	35.5	31.5	25.3	28.9	24.5	0.01	0.12	15.9	18.3 ± 15.1
Temperature$_{surface}$ (°C)	upper	21.6	24.0	19.6	24.0	25.2	23.3	25.3	26.1	21.6	15.0	17.5	21.1	22.0 ± 3.35
	middle	23.0	24.2	20.6	26.3	26.5	24.2	26.0	27.0	24.0	16.4	17.8	22.0	23.2 ± 3.43
Temperature$_{bottom}$ (°C)	upper	21.6	24.0	19.6	25.2	25.2	23.3	26.1	25.9	21.0	14.8	17.5	21.1	22.1 ± 3.54
	middle	23.2	24.8	20.6	27.4	26.3	24.2	26.7	27.7	24.5	15.3	17.8	21.7	23.4 ± 3.88
TOC (%)	upper	7.4	5.1	8.9	12.4	9.7	8.9	10.8	17.1	19.5	20.0	13.5	9.2	11.9 ± 4.79
	middle	31	31	35	35	32	34	31	30	29	29	35	36	32.4 ± 2.70
Phosphate (mg/L)	upper	0.4	0.4	0.3	0.6	0.8	0.6	0.4	0.4	0.2	0.2	0.4	0	0.39 ± 0.21
	middle	0.2	0.2	0.2	0.0	0.4	0.0	0.0	0.2	0.2	0.0	0.0	0.1	0.13 ± 0.13
Oxygen$_{bottom}$ (mg/L)	upper	7.2	10.3	9.1	5.0	9.1	10.0	8.9	8.4	8.2	8.7	10.1	10.0	8.76 ± 1.49
	middle	7.5	8.7	8.3	4.8	7.4	7.5	7.7	8.5	8.0	9.0	9.0	8.2	7.88 ± 1.11
Velocity (m/sec)	upper	0.16	0.06	0.09	0.03	0.03	0.04	0.02	0.02	0.03	0.4	0.4	0.09	0.11 ± 0.14
	middle	0.02	0.02	0.02	0.02	0.04	0.04	0.02	0.02	0.02	0.02	0.02	0.02	0.02 ± 0.01

The site within the middle reaches of the estuary is characterised by a higher stratification and a wider range of salinities (0.01 to 8.17 ppt at the surface and 0.01 to 42.7 ppt at the bottom) compared to the site within the upper reaches (0.01 to 0.45 ppt at the surface and 0.01 to 8.17 ppt at the bottom). Furthermore, the site within the middle reaches showed a significantly higher temperature ($p < 0.001$) and TOC ($p < 0.001$) and significantly lower phosphate concentration ($p = 0.002$), dissolved oxygen ($p = 0.007$) and velocity ($p = 0.05$) than the site situated in the upper reaches (Table 4.1).

Particle-bound pesticides like CPF, total END, p,p-DDE, cypermethrin (CPM), prothiofos (PTF) and fenvalerate (FENV) were frequently detected at both sites. In general, the upper reaches of the estuary showed higher average concentrations of CPF (7.5 ± 10.9 µg/kg), total END (23.2 ± 53.5 µg/kg), p,p-DDE (24.4 ± 26.9 µg/kg) and CPM (3.4 ± 3.5 µg/kg) than in the middle reaches (3.7 ± 4.6 µg/kg, 2.1 ± 3.5 µg/kg, 22.1 ± 38.4 µg/kg, 2.3 ± 2.8 µg/kg, respectively). Only PTF showed higher concentrations within the middle reaches (6.2 ± 6.3 µg/kg) compared to the upper reaches (5.7 ± 3.4 µg/kg).

In general, organochlorines (total END and p,p-DDE) showed the highest concentrations, followed by organophosphates (CPF and PTF), whereas pyrethroids (CPM and FENV) showed the lowest concentrations. The midstream site has higher levels of TOC and salinity and lower levels of ortho-phosphate (Figure 4.2) as well as lower levels of pesticides, mainly total END and CPF (Figure 4.3).

4.3.2.2) First runoff event

The first runoff event on the 15th January resulted in little alteration of most environmental variables (Table 4.1, Figure 4.2). Within the upper reaches of the estuary only the bottom salinity decreased rapidly (from 10.1 to 0.12 ppt), whereas the other environmental variables (including velocity) did not change significantly (Table 4.1). Within the middle reaches the bottom salinity decreased due to the runoff event but still remained at high levels (35.5 ppt).

Since TSS (Figure 4.5) and discharge from the upstream catchment (Figure 4.1) increased on the 16th January 2003, one day after the rainfall event, it is assumed that the particle-bound pesticide concentrations peaked only a day after the rainfall event. Within the upper reaches, a drastic increase was measured in CPF (from 0 to 27.1 µg/kg), CPM (from 5.2 to 12.3 µg/kg) and total END (from 5.5 to 191 µg/kg), whereas p,p-DDE decreased from 61.3 to 0 µg/kg. Thus it is important to stress out that the first runoff event needed to be distinguished between day of environmental change (15th January) and day of pesticide increase (16th January)

Spatial and temporal variability

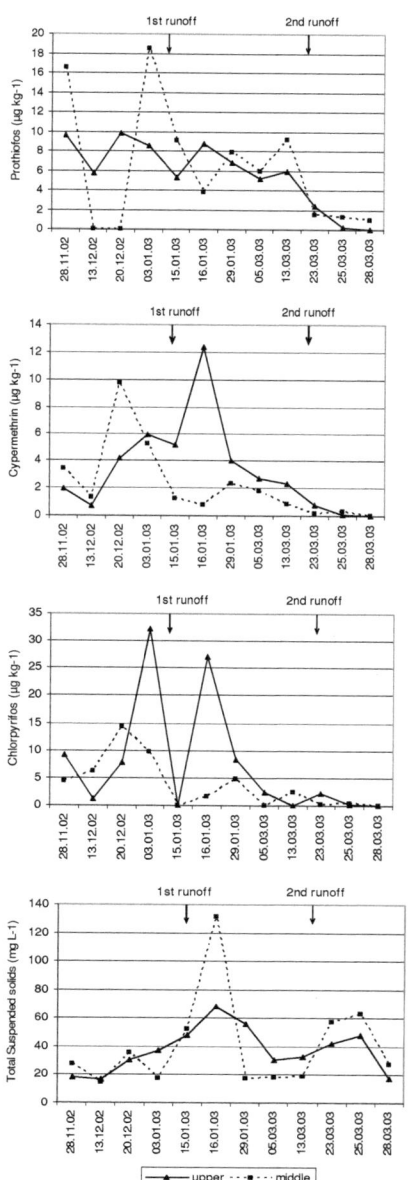
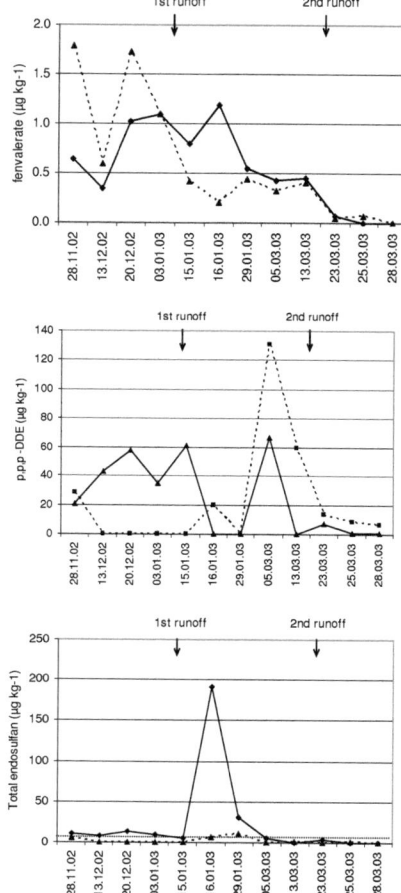

Figure 4.5 Temporal variation of particles associated Pesticides within the upper and middle reaches of the Lourens River estuary during the study period from November 2002 to March 2003, with emphasis on the two runoff events, indicated by the two arrows.

Due to the increase in TSS from 52.2 mg/L to 131.3 mg/L within the middle reaches from the 15th to the 16th January, it might be assumed that the particle-bound pesticides should also have reached the downstream site by that time. However, only a slight increase in CPF (from 0 to 1.6 µg/kg), total END (from 0 to 6.2 µg/kg) and p,p-DDE (from 0 to 19.4 µg/kg) was recorded at the downstream site. On the 16th January (after the runoff event) less pesticide concentrations were found at the downstream site compared to the upstream one (94% less CPF, 93% less CPM, 83% FENV and 97% for total END). Only p p-DDE showed higher concentration within the middle reaches with 19.4 µg/kg.

4.3.2.3) Second runoff event

The second, much stronger, runoff event (rainfall 30.2 mm; 23rd March 2003) resulted in drastic changes in environmental variables (Table 4.1, Figure 4.2), but no alteration in particle-bound pesticide concentrations at both sites (Figures 4.3 and 4.5). Salinity decreased to 0.01 at both sites (Table 4.1), temperature decreased by approximately 6 °C at the upstream site and by 9 °C at the downstream site. The flow increased drastically from 0.03 to 0.4 m/sec within the upper reaches. In contrast to the situation during the first runoff event, high TSS concentrations were measured on the day of the runoff event, compared to the following day. However, no significant change in particle-bound pesticide concentration was measured during the second runoff event in association with the TSS increase at both sites.

4.3.3) Response indicators

4.3.3.1) General spatial variability

A total of 27 taxa (including epi-benthic and hyper-benthic meiofauna and demersal zooplankton) were found all in all within the upper reaches of the estuary, whereas only 20 taxa were identified within the middle reaches. Both communities were dominated similarly by the copepods *Halicyclops*, *Mesochra* and *Canthocamptus*. Combined, they accounted for 61% and 79% of the total abundance in the upper and middle reaches, respectively. Some of the taxa absent from the middle reaches were freshwater taxa like the cladocera *Dunhedevia* and *Daphnia* and the macroinvertebrate families Baetidae, Hydropsychidae and Perlidae. Generally, the benthic community within the upper reaches was characterised by lower abundance (1,537 ± 1,541 Ind/m^3), higher number of taxa (8.1 ± 2.9) and higher Shannon Wiener Diversity Index (1.4 ± 0.31) than the benthic community in the middle reaches with 7,775 ± 12,503 Ind/m^3, 5.4 ± 2.4 taxa, and a H´ of 0.61 ± 0.25 (Table 4.2).

The spatial differences were characterised by a higher abundance of estuarine species, mainly *Mesochra* and *Pseudodiaptomus* at the middle reaches, whereas the upper reaches were more characterised by freshwater taxa like *Dunhevedia* and *Thermocyclops*. The harpacticoid *Mesochra* contributed the most to the spatial differences in community structure (Figure 4.4).

Table 4.2. Total abundance, number of taxa and Shannon Wiener Diversity Index during the study period within the middle and upper reaches of the Lourens River estuary.

	Total abundances/ m^3		No of taxa/ m^3		Shannon-Wiener Index	
	upper	middle	upper	middle	upper	middle
28.11.02	258	191	5	4	0.86	0.62
13.12.02	1231	5044	7	3	1.0	0.29
20.12.02	316	124	7	1	1.6	0.13
03.01.03	516	15475	7	6	1.8	0.55
15.01.03 [1a]	4360	12080	14	7	1.5	0.67
16.01.03 [1b]	2396	43440	9	8	1.8	0.68
29.01.03	2667	11782	11	8	1.5	0.77
05.03.03	480	760	5	4	1.5	0.50
13.03.03	4391	1848	12	7	1.6	0.83
23.03.03 [2]	837	1991	8	8	1.6	0.96
25.03.03	516	377	7	6	1.1	0.89
28.03.03	484	186	5	3	1.0	0.45
average	1538 ±1542	7775 ± 12502	8.1 ± 2.9	5.4 ± 2.4	1.4 ± 0.31	0.61 ± 0.25

1a 1st runoff event (changes of environmental variables)
1b 1st runoff event (change of particle-bound pesticide concentration)
2 2nd runoff event

4.3.3.2) First runoff event

During the rainfall event on the 15th of January, the upper site was characterised by an increase in abundance (from 516 to 4360 Ind/m^3), number of taxa (from 7 to 14) and a decrease in Shannon Wiener Diversity Index (from 1.8 to 1.5). The increase in taxa from 3rd to 15th January was mainly due to the increase in freshwater species like *Dunhevedia*, *Themocyclops*, *Ceriodaphnia*, Perlidae and Baetidae (Figure 4.6). The middle reaches hardly showed any change due to the lack of freshwater input (Table 4.2).

One day after the rainfall event on 16th January, when the particle-bound pesticide concentrations increased in the upper reaches, there was a decrease in abundance (4,360 to 2,396 Ind/m^3), no of taxa (from 14 to 9) due to the decrease in freshwater species but also due to a decrease in *Mesochra* and *Halicyclops*.

Within the middle reaches, however, only the abundance increased rapidly from 12,080 to 43,440 Ind/m^3 (Table 4.2) mainly due to the increase of two taxa: *Canthocamptus* and *Mesochra* (Figure 4.6).The PRC shows the highest deviation in community structure between the two sites one day after the first rainfall event (16th January), mainly due to *Mesochra* (Figure 4.4).

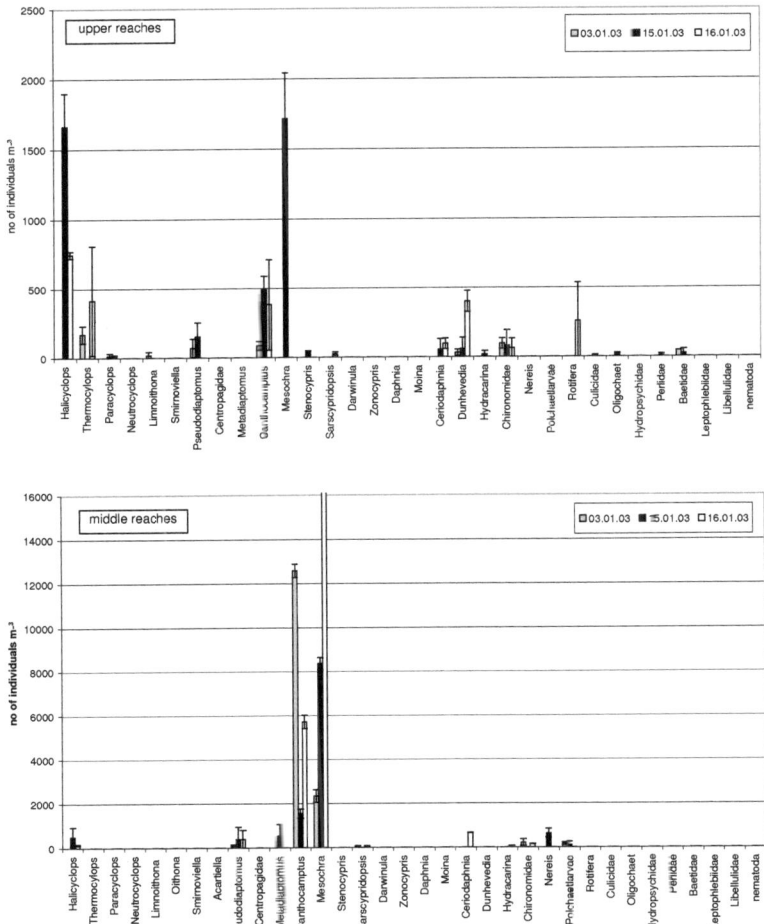

Figure 4.6. Response of benthic taxa to the first runoff events within the upper and the middle reaches of the Lourens River estuary during the study period from November 2002 to March 2003.

3.3.3) 2nd runoff event

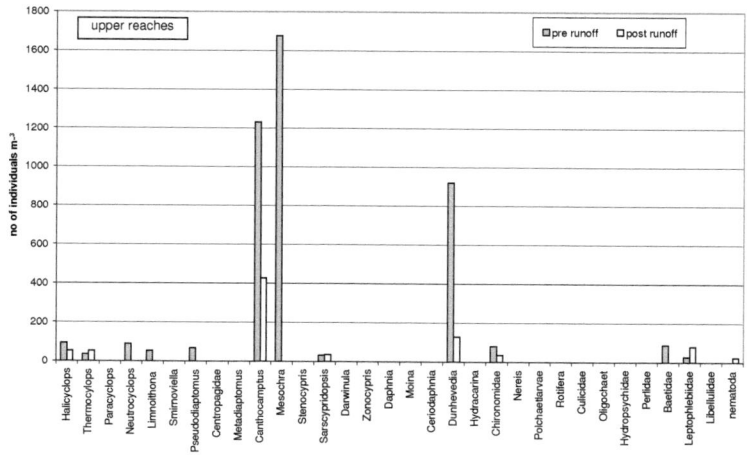

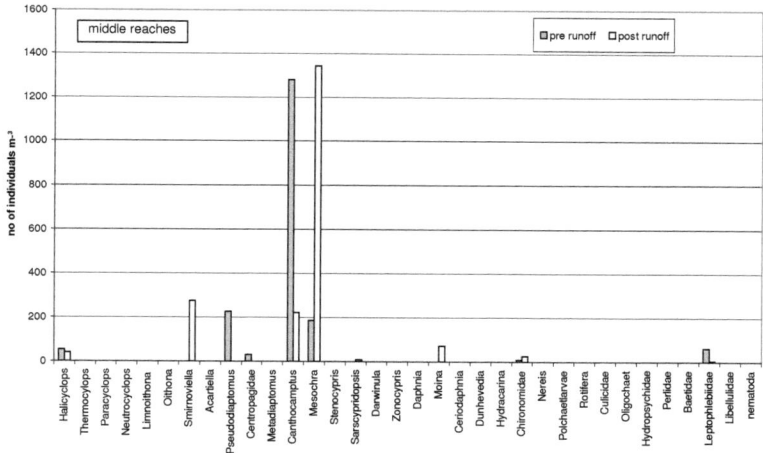

Figure 4.7. Response of benthic taxa to the second runoff events within the upper and the middle reaches of the Lourens River estuary during the study period from November 2002 to March 2003.

As a result of the second runoff event, the abundance and number of taxa decreased in the upper reaches from 4,391 to 837 Ind/m^3 and from 12 to 8 taxa, mainly due to a decrease in numbers of *Mesochra* and *Canthocamptus* (Table 4.2; Figure 4.7), and remained constant in constant in the middle reaches.

The Shannon Diversity Index remained constant in the upper reaches and increased to the maximum of 0.96 within the middle reaches. The PRC showed a higher spatial similarity in community structure after the second runoff event compared to the first runoff event (Figure 4.4).

4.4 DISCUSSION

4.4.1) Environmental variables and particle-bound pesticides

In general, lower pesticide concentrations (mainly total END and CPF) were detected in the middle reaches than in the upper reaches. The spatial difference in pesticide concentrations was particularly evident during the high input of pesticides measured at the upper reaches one day after the first runoff event, as successfully shown by the PRC analysis (Figure 4.3).

The high input upstream was proven to be linked to an increase in TSS and flow (further upstream), whereas the slight increase in CPF, PTF and total END within the middle reaches was not linked to the drastic increase in TSS and the increase in discharge from further upstream. The reduced pesticide concentrations measured further downstream in the estuary may have various causes. Macrophytes have been identified as an important factor in sorption of pesticides (Schulz & Peall, 2001), such as CPF (Moore et al., 2002) or azinphos-methyl (Dabrowski et al., 2006). Since the stretch between the two sites is covered by 40% of *Juncus capensis* and other macrophytes, reduction through sorption is a plausible mechanism for the observed decrease in pesticide concentration.

The midstream site features a strong marine influence, as implied by the high salinity measurements. A seaward reduction in chemical concentration is frequently observed due to the long term mixing of contaminated river particles with less contaminated marine particles (Turner & Millward, 2002). The midstream site showed higher TSS and p,p-DDE concentrations, which probably originated from the sites itself. Therefore dilution of pesticides at the midstream site by particles of marine origin may explain lower concentrations of pesticides compared to the upstream site.

The second, much stronger runoff event, which took place during low-spraying season, did not feature any increase in pesticide concentration. There might be several explanations for this observation. Pesticides are frequently applied between August and February before fruit harvest (Dabrowski *et al.*, 2002), with the bulk applied from November to January (Schulz, 2001). Pesticides could have been already degraded on the field (Antonius & Byers, 1997) or washed off, and therefore no new pesticide input would have taken place during the runoff event. The high velocity as a result of the runoff event could have been too high, for particles to be able to settle down into the sediment sampler. The increase in freshwater volume as a result of the runoff event could have significantly diluted the concentration of particle-bound pesticides.

4.4.2) Response indicators

Hartwell & Clafin (2005) assigned a Shannon Wiener Index for macrobenthos < 1.5 a score of only 1, between 1.5 and 2.4 a score of three and indices above 2.4 a score of 5. Bollmohr (unpublished data) measured a diversity index from 1.1 to 2.0 in a less disturbed temporarily open estuary 20km south of the Lourens River estuary. Therefore the Shannon Wiener Index from 0.13 to 1.79 recorded at both sites within the Lourens River estuary can be categorised as low. Most of the measurements fell below score 1. The low diversity scores could be a result of i. the almost closed status of the estuary with no exchange with the seawater and lack of marine species (Kibirige & Persissinotto, 2003); or ii. a chronically stressed environment due to frequent exposure to pesticides (Figure 4.5).

However diversity indices provide little information on the state of pollution in an estuary, since estuaries are strongly influenced by salinity (Hartwell & Clafin's, 2005). The highest number of species occurs either in fresh or marine water. Since much fewer species occur at intermediate salinities, such as in estuaries, comparing diversity spatially along an estuary (as in this study) is very difficult. Furthermore, it can be positively influenced by the short term transport of freshwater species after a runoff event. The higher diversity in the upper reaches is probably caused by the input of freshwater species (Figure 4.6). These species did not reach the middle reaches or if they did, they were not able to survive in the more saline environment. The Shannon Diversity Index is sensitive to low species abundance and might not appropriately characterise the effects of pollution on estuarine biodiversity. In addition, an observed artefact of the Shannon Diversity Index is that it could be artificially increased if a species exhibited a decrease in abundance.

Bollmohr *et al.*, (2009b) showed that the survival rate of a common species in the Lourens River estuary (*Mesochra parva*) can be differentiated between different salinities and different exposure patterns, as well as between male and female organisms. Male organisms are more sensitive to pesticide exposure and salinity decrease. Therefore the abundance of male *M. parva* could decrease as a result of pesticide exposure while the Diversity Index increases. High levels of organic enrichment decrease species diversity and increases abundances of species (Gee & Warwick, 1985). The higher diversity could also be a result of lower organic content at the upstream site compared to the midstream one.

The PRC analysis provides information, not provided by general community indices, which could be used to determine pesticide effects on meiobenthos communities in estuaries. Temporal variability in the dataset explained more variation in community structure, environmental variables and pesticide concentrations than spatial variability. This is explained by the extremely high variability in temporarily open estuaries (Kibirige & Perissinotto, 2003) as a result of freshwater input and season. An increase in spatial differences in community structure was revealed by PRC for both the first and second runoff events. The spatial difference in community structure during the first runoff event was probably caused by pesticide concentrations and the difference during the second event was caused by environmental variables. The effects of pesticides can thus be successfully distinguished from the impact of natural variables by the PRC analysis.

The pesticides exhibiting the greatest impact in the system are total END and CPF. This finding is supported by species sensitivity distribution studies by Bollmohr et al., (2007), which showed that END and CPF pose the highest risk towards freshwater and marine communities in the Lourens River estuary. Only a few studies have investigated changes in estuarine community structures due to pesticide exposure. Results of the study done by deLorenzo et al. (1999) suggested that exposure to agricultural pesticides can lead to both functional and structural changes in the estuarine microbial food web. Endosulfan primarily reduced bacterial abundance, the number of cyanobacteria and phototrophic biomass. Chlorpyrifos decreased the abundance of protozoan grazers. Chandler & Green (2001) found a decrease in production of *Amphiascus tenuiremis* copepodit and naupliar during CPF exposure of 11-22 µg kg^{-1}. A predicted sediment quality criterion for a maximum safe concentration of chlorpyrifos was calculated to be 31.2 µg/kg. Leonard et al. (2001) determined a 10-d NOEC of 42 µg/kg for sediment associated END for the epibenthic mayfly *Jappa kutera*. Chandler & Scott (1991) suggested that polychaete (*Streblospio benedicti*) populations in the field may be strongly depressed by sediment endosulfan concentration of 50 µg/kg. The pesticide concentrations at which effects should be observed in estuarine systems suggested by various authors (Chandler & Scott, 1991; Chandler et al., 1997; Chandler & Green, 2001; Bollmohr et al., 2007) are exceeded during the first runoff event in this study.

The natural variables contributing most to the differences between the two sites are TOC and salinity, with higher values of each of these variables found at the midstream site. This supports the findings of previous studies (Austen, 1989; Perissinotto et al., 2002). However, temperature and grain size, which are often stated as important variables contributing to seasonal variability in estuarine communities (Nozais et al., 2005), show a low importance in the Lourens River estuary.

Species contributing to the temporal and spatial difference were mainly species belonging to the harpacticoid genus *Mesochra*, mainly driven by the environmental variables TOC and salinity and the particle-bound pesticides total END and CPF (Figures 4.2, 4.3 and 4.4). Bollmohr et al. (2009b) investigated the interactions between adaptations to salinity changes, CPF exposure and salinity decrease on the survival rate of *Mesochra parva*. They suggested a synergistic effect between the latter two and a detrimental effect of CPF exposure at concentrations of only 6 µg/kg, being exceeded by 42% of the samples from the upper reaches. This supports the finding of this study, indicating *Mesochra* as one of the species most sensitive to salinity and CPF exposure.

4.5 CONCLUSION

The temporarily open estuarine system is characterised by a higher temporal than spatial variability of environmental variables, particle bound pesticides and community structure. Due to the long term exposure of pesticides with toxicologically relevant concentrations, it remains difficult to determine any acute effect towards the dynamic of already adapted meiobenthos community. Furthermore, the generally low biodiversity index suggests a rather chronic than acute effect on the system. However, within the spatial variability (between upper and middle reaches) the authors were able to detect a link between endosulfan, chlorpyrifos exposure, TOC and salinity and community change by comparing two runoff events.

CHAPTER 5

INTERACTIVE EFFECT OF SALINITY DECREASE, SALINITY ADAPTATION AND CHLORPYRIFOS EXPOSURE ON AN ESTUARINE HARPACTICOID COPEPOD, MESOCHRA PARVA, IN SOUTH AFRICA

Published

S. Bollmohr, R. Schulz, T. Hahn

(Ecotoxicology and Environmental Safety, in press)

5.1 INTRODUCTION

Salinity is one of the main environmental factors controlling species distribution, abundance and sex-ratio in marine, and particularly in estuarine organisms (Willmer *et al.*, 2000). Estuarine organisms are mainly euryhaline, experiencing periodical acute changes of the ionic milieu of their habitat, and have evolved regulatory mechanisms to maintain vital processes. Such a capacity for regulation is often connected to a period of adaptation (Stucchi-Zucchi and Salomão, 1998; Anger, 1996). A number of studies have addressed the relative effects of contaminants on estuarine species (Di Pinto et al., 1993; Chandler *et al.*, 1997; Chandler and Green, 2001), but rarely have these effects been measured in the context of toxicant/ salinity stress interaction (Staton *et al.*, 2002).

The experimental approach presented here was based on typical field conditions observed in the temporarily open Lourens River estuary, Western Cape, South Africa. Natural conditions in this estuary are characterised by fluctuations in freshwater input and tidal exchange. Due to an increase in water abstraction in South Africa (Scharler and Baird, 2003), the freshwater input during the summer season (November to February) is decreasing, which results in higher and more constant salinities within the estuary. Under these conditions estuarine organisms may not be able to develop their adaptive capacities due to increasingly constant salinity conditions.

In the Western Cape region most pesticides are applied during the summer season, thus most of these are expected to enter the estuary during this period. A runoff event after rainfall will transport considerable amounts of these pesticides downstream into the estuary (Bollmohr *et al.*, 2007). For the estuarine fauna this would result in a combination of effects, mainly (1) salinity decrease due to dilution with freshwater and (2) increase in pesticide exposure. The organophosphate pesticide chlorpyrifos (CPF) was measured in high concentrations upstream the Lourens River with 30 µg/kg during summer 1999 and 69 µg/kg during a runoff event (Dabrowski *et al.*, 2002). Bollmohr *et al.* (2007) reported that, out of nine pesticides detected in the Lourens River, CPF poses the highest risk to estuarine organisms in the Lourens River estuary, based on sediment and water concentrations and toxicity data. The majority of CPF does not remain in aqueous solution/suspension but binds to the organic and clay fractions of sediments and deposits in estuarine ecosystems, due to its high K_{OC} value (organic carbon/water partitioning coefficient) of 5,011 L/kg (Sabljic *et al.* 1995). In the upper 1 to 2 cm oxic zone of muddy sediments live more than 98% of all meiobenthic copepods. These animals are an important link in the marine food web and serve as a food source for many larval and juvenile penaeid shrimp and fish species (Dall et al., 1990; Hicks and Coull, 1983; McCall and Fleeger, 1995). They also play important roles in marine ecological and physicochemical processes, such as cycling and remineralisation of organic carbon (Hicks and Coull, 1983; Coull, 1990).

Due to their habitat, meiobenthic copepods may be particularly affected by exposure to particle-associated pesticides (Chandler et al., 1997). Green et al. (1996) found that CPF is highly toxic to the copepod *Amphiascus tenuiremis* (96-h LC50 = 66 µg/kg sediment) with a greater sensitivity in early (i.e., nauplius) life stages (96-h LC50 = 40 µg/kg sediment) (Green et al., 1996).

In this study we performed a sediment toxicity test with *Mesochra parva* (Thompson 1946), a meiobenthic harpacticoid copepod species, which is seasonally abundant in the Lourens River estuary. The aim was to determine the interactive effect of pre-adaptation to varying salinities, salinity decrease, and CPF exposure on the survival rate of females and males. We hypothesised that pre-adaptation to fluctuating salinities would lead to enhanced survival when exposed to a combination of CPF exposure and hypoosmotic stress. Furthermore the response of *M parva* to the different tested variables with the laboratory test was compared with the response in the Lourens River estuary.

5.2   MATERIALS & METHODS

5.2.1) Preparation of clean/sediment substrates

Surface sediments (0-2cm) were collected from a reference site, the Rooiels estuary (Western Cape, South Africa; 34°18'S and 18°49'E), of which catchment is characterised by natural fynbos vegetation, and thus no pesticide contamination is expected to be found. The sediment was sieved through a 63-µm sieve and washed with distilled water as described by Chandler and Green (1996). Afterwards the sediment was dried for 24 h at 180°C. Sediment organic carbon (f_{OC}) was 1.35 ± 0.27% and average grain size of the silty-clay sediment was 226 ± 15 µm. The water used in the bioassay contained a mixture of seawater, collected at the Rooiels estuary mouth and freshwater, collected from an unnamed pristine Table Mountain stream. Both media were mixed to achieve salinities of 3 parts-per-thousand (ppt) and 15 ppt for the different test runs. The salinities were chosen based on average salinities in the Lourens estuary (Fig. 5.4a).

5.2.2) Sediment CPF spiking

Chlorpyrifos (O,O-diethyl O- [3,5,6-trichloro-2-pyridyl] phophorothioate, CPF) was applied as emulsifiable concentrate (DURSBAN, South Africa) containing 480 g/L of active ingredient. The emulsifiable concentrate, rather than the pure substance was used in order to approximate the conditions in the field. Apart from the active ingredient, the emulsifiable concentrate contains emulsifier (composition not reported) and water. All concentrations given here refer to the active ingredient. A stock solution was prepared by dissolving CPF in acetone carrier. From this, an adequate volume was added under mixing to 100 g sediment to deliver the desired CPF concentration of 8 µg/kg sediment dry weight. This test concentration was chosen because it resulted in approximately 50% mortality in preliminary range finding tests (Bollmohr, unpublished data). Control treatments received an equal volume of pure acetone. After spiking, all sediments were homogenised and left for an additional 12 h at 20°C before use.

5.2.3) Copepod collection

Mesochra parva harpacticoid copepods were collected from the Rooiels estuary by sampling the interface between sediment and water column with a plankton net (Ø 30 cm, mesh size 80 µm). The copepods were concentrated by sieving through a 500-µm sieve and were retained by a 125-µm sieve. During sampling (14.02.2003) the salinity in the estuary was 6.4 ppt at the surface and 30.5 ppt at the bottom. The sampled material was immediately transported into the laboratory. Animals were isolated from the sample and transferred into two different beakers, each filled with Rooiels sediment and water with a salinity of 15 ppt. An initial 24 h test treatment demonstrated a mortality of less than 10% following direct transfer of copepods to 3 ppt or 15 ppt.

5.2.4) Bioassay design

After collection of the copepods, 800 individuals (400 males + 400 females) were counted into each of two different culture beakers. Both populations were cultured for 28 d at 20°C to establish a new laboratory generation. During this time period one culture was pre-adapted to daily salinity changes between 15 ppt and 3 ppt by exchange of the medium, whereas the other culture was reared under constant salinity conditions (daily water change at a constant salinity of 15 ppt) (Figure 5.1)

Interactive effects

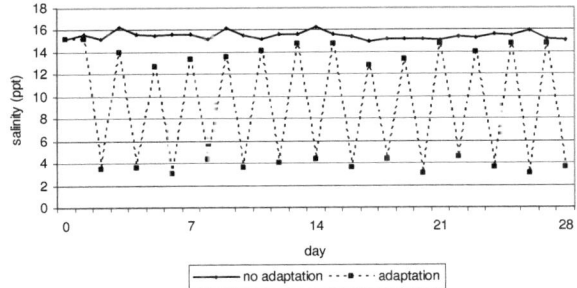

Figure 5.1 Measured salinity fluctuation during the adaptation period of 28 days.

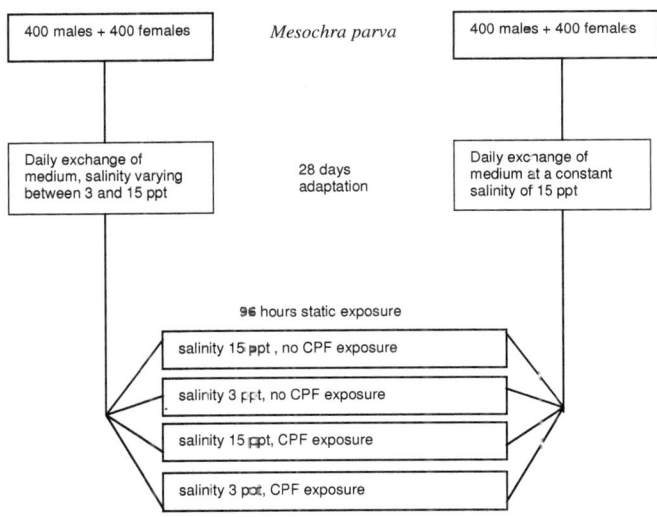

Figure 5.2 Illustration of the full factorial experimental design to evaluate the combined influence of adaptation to fluctuating salinity, salinity decrease and chlorpyrifos exposure in a 96-hour sediment toxicity test with *Mesochra parva*

A 96-h sediment toxicity test with 8 µg/kg CPF exposure was then conducted to analyse the survival rate of male and female *Mesochra parva* separately under the experimental variables. Six replicate treatments with 15 individuals each (female and male) were set up, containing 15 g sediment and 100 mL medium for each combination of experimental variables. Survivors were defined as those individuals with normal, unimpaired swimming ability. The experimental setup is illustrated by Figure 5.2.

5.2.5) Water-quality and chlorpyrifos analysis

Physico-chemical water parameters were measured with test kits from Macherey & Nagel (Düren, Germany) and electronic meters from Wissenschaftliche Technische Werkstätten (Weilheim, Germany). All parameters were measured in two parallel treatments at the beginning and at the end of the experiment. CPF concentrations were measured in the sediment and in the water at the beginning and at the end of the experiment. Due to the small amount of water and sediment per treatment, the six replicates (per treatment) were combined for analysis at the Forensic Chemistry Laboratory (Department of Health, Cape Town, South Africa). A detailed description of the extraction and analytical procedure is given by Dabrowski *et al.* (2002). Briefly, water samples were solid-phase extracted using Chromabond® C18 columns (Macherey-Nagel, Düren, Germany) which had been previously conditioned with 6 mL methanol and 6 mL water. The columns were air dried for 30 min and kept at -18 °C until analysis. The sediment samples were placed in a glass jar and were kept at -18 °C until analysis. Analyses were performed by gas chromatography using a Hewlett-Packard model 5890 (Hewlett-Packard, Avondale, PA, USA), fitted with standard Hewlett-Packard electron-capture, nitrogen-phosphorus, and flame-photometric detectors. The detection limits were 0.01 µg/L and 0.1 µg/kg dry weight for water and sediment samples, respectively, and spike recovery rates were between 79% and 106%.

5.2.6) Field measurements

Conductivity and temperature were measured at the bottom to determine stratification trends within the estuary, using a conductivity meter LF 91 from Wissenschäftliche Technische Werkstaetten, Weilheim, Germany. The salinity was obtained by converting the conductivity values into salinity using an automatic converter (Fofonoff and Millard, 1983). Samples of insecticides associated with suspended sediments were accumulated continuously by a suspended-particle sampler (Liess *et al.*, 1996), from which they were collected at 14-days intervals at both estuaries. The suspended-particle sampler consisted of a plastic container (500 mL) with a screw-on lid containing a hole (2 cm in

diameter), which contained an open glass jar that was stored directly under the hole in the lid. The samplers were attached to a metal stake and were stored on the riverbed.
A different sampling approach was applied to reduce the sampling error caused by small-scale variability in sediment and to sample the organisms living in the flocculent layer between sediment and water column. Therefore a stretch of 10 m flocculent layer was sampled just above the bottom sediment using a zooplankton net with 300ym mesh-size. Four replicates were sampled and within each replicate the zooplankton/ meiofauna was identified to the lowest taxonomic unit possible, and the abundance was determined based on six sub-samples.

5.2.7) Statistical analysis

Differences in survival rate between the treatments were analyzed using a three-factorial analysis of variance (3-way ANOVA) with (1) adaptation to varying salinities, (2) salinity, and (3) CPF exposure as effect variables, and survival rate as dependent variable. Survival rate data were Square-root-arcsine transformed before performing the 3-way ANOVA to fulfil criteria of normal distribution and variance homogeneity. Tukey's HSD test was performed to describe differences between the various treatments. Differences between survival rates of female and male harpacticoid copepods exposed to the same combination of effect variables (CPF, salinity, adaptation) were analysed using Student's t-test.

5.3 RESULTS

5.3.1) Water quality and CPF analyses

Water chemical parameters were analysed initially and after termination of the sediment toxicity experiment. Mean values (± standard error, n=2) of pH, conductivity, salinity, temperature, oxygen and nutrients (nitrate, nitrite, ammonia, phosphate) are shown in Table 5.1. There were no significant differences in between initial and final values and between the two different salinities. The results of the CPF analyses are presented in Table 5.2.

Table 5.1 General water and sediment quality measurement (n=2) taken at the beginning of the experiment and after 96h.

		3ppt		15ppt	
		0h	96h	0h	96h
water					
pH		6.9 ± 0.1	7.3 ± 0.1	7.2 ± 0.1	7.3 ± 0.1
Conductivity	mS/cm	5.0 ± 0.1	5.7 ± 0.2	22.4 ± 0.1	23.2 ± 0.2
Salinity	ppt	3.0 ± 0.0	3.1±0.2	15.0 ± 0.0	15.4±0.2
Temperature	°C	20.0 ± 0.1	20.2 ± 0.0	20.0 ± 0.1	20.2 ± 0.1
Oxygen	%	94.0 ± 0.9	92.7 ± 1.2	93.7 ± 0.8	93.0 ± 0.9
Nitrate	mg/l	0.33 ± 0.26	0.02 ± 0.03	0.08 ± 0.20	0.02 ± 0.03
Nitrite	mg/l	0.00 ± 0.00	0.00 ± 0.00	0.00 ± 0.00	0.00 ± 0.00
Ammonia	mg/l	0.00 ± 0.00	0.00 ± 0.00	0.00 ± 0.00	0.00 ± 0.00
Phosphate	mg/l	0.03 ± 0.02	0.03 ± 0.03	0.02 ± 0.03	0.03 ± 0.03
sediment					
TOC	%	1.32	1.35	1.35	1.22
Median grain size	µm	215	211	241	237

Measured initial sediment concentrations were an average of 84 ± 3% of the nominal concentrations, and of 67 ± 5% after 96 hours. The 15 ppt-salinity treatment showed higher CPF concentrations in the sediment (5.60 – 5.76 µg/kg), but lower concentrations in water (0.8 – 1.0 µg/L) compared to the 3 ppt treatment (5.00 – 5.13 µg/kg and 1.5 µg/L respectively). In sediment and water of the control vessels no CPF was detected at the detection limits of 0.1 µg/kg and 0.01 µg/L, respectively.

Table 5.2 Measured chlorpyrifos concentrations in spiked sediment and overlying water in the various experimental treatments.

	Measured concentration (0 h)	Measured concentration (96 h)
	Sediment (µg/kg)	
3 ppt salinity, males	6.38	5.00
3 ppt salinity, females	6.69	5.13
15 ppt salinity, males	6.84	5.60
15 ppt salinity, females	7.00	5.76
	Water (µg/L)	
3 ppt salinity, males	0.3	1.5
3 ppt salinity, females	0.3	1.5
15 ppt salinity, males	ND	1.0
15 ppt salinity, females	ND	0.8

ND = not detected; limit of detection = 0.1 µg/kg of CPF; 0.01µg/L

[a] no CPF was detected in setups without experimental CPF spiking

5.3.2) Copepod survival in relation to experimental factors

Results of the three-factorial ANOVA (Table 5.3) showed that pre-adaptation to fluctuating salinities ($p=0.02$; $p=0.001$), salinity decrease ($p=0.035$; $p<0.001$) and CPF exposure ($p<0.001$; $p<0.001$), all had a significant impact on the survival rate of female and male *M. parva* harpacticoid copepods. However, statistical significance for interactive effects of the experimental factors were detected only for salinity x CPF exposure and salinity x adaptation, but not for CPF exposure x adaptation and not for the combination of all three factors in both, male and female harpacticoids (Table 5.3).

Table 5.3. Three factorial analysis of variance for the effects of adaptation to varying salinities (adapt), salinity (sal) and CPF exposure (cpf) on survival of female and male *Mesochra parva* (squareroot arcsin transformed) [a]

3-way ANOVA Males					
effect	df	SS	MS	F	p
adap	1	0.31	0.31	12.9	0.001
sal	1	0.81	0.81	33.6	0.000
cpf	1	12.49	12.49	519.5	0.000
adapt*sal	1	0.40	0.40	16.5	0.000
adapt*cpf	1	0.01	0.01	0.2	0.631
sal*cpf	1	0.38	0.38	15.7	0.000
adapt* sal *cpf	1	0.08	0.08	3.3	0.077
error	40	0.962	0.024		
total	47	15.433	0.328		

3-way ANOVA Females					
effect	df	SS	MS	F	p
adapt	1	0.113	0.113	5.6	0.023
sal	1	0.096	0.096	4.8	0.035
cpf	1	3.857	3.857	192.8	0.000
adapt*sal	1	0.116	0.116	5.8	0.021
adapt*cpf	1	0.011	0.011	0.6	0.458
sal*cpf	1	0.174	0.174	8.7	0.005
adapt* sal *cpf	1	0.000	0.000	0.0	0.894
error	40	0.800	0.020		
total	47	5.167	0.110		

[a] df = degrees of freedom; F = likelihood ratio; MS = mean of squares, SS = sum of squares, p = probability

Statistical differences between the various experimental treatments were analysed using Tukey's post-hoc procedure (Figure 5.3). Without CPF addition, exposure of males to a salinity of 3 ppt resulted in significantly reduced survival of animals unadapted to daily salinity fluctuation (46 ± 7.8%), when compared to pre-adapted males at both salinities (3 ppt: 86 ± 5.0%; 15 ppt: 93 ± 6.0%) and to unadapted males at 15 ppt (97 ± 3.7%). In females this effect was more ambiguous. Similarly to the male results, unadapted at 3 ppt females survived at a lower rate (81 ± 7.8%) than pre-adapted females (91 ± 5.4%), but statistical significance was not determined. However, survival of the unadapted female group at 3 ppt was significantly lower (81 ± 7.8%) than that of unadapted females (97 ± 3.7%) at 15 ppt salinity.

The above analysis reveals that male and female survival in groups exposed to CPF is significantly lower than in groups without CPF. But, unlike the non-CPF groups, in groups with CPF exposure no statistically significant differences related to adaptation and/or salinity decrease were detected.

When looking at the differences between groups pre-adapted and non-adapted to daily salinity changes at 3 ppt, however, it is striking that male and female survival of non-adapted copepods was always lower than that of pre-adapted copepods. In males $0 \pm 0\%$ survival was found in non-adapted animals, while in the pre-adapted group $8.9 \pm 10\%$ survivors were recorded. Also, survival rates recorded at 15 ppt were similar for pre-adapted ($7.8 \pm 9.8\%$) and non-adapted males ($6.7 \pm 8.4\%$).

In females, survival of non-adapted copepods at 3 ppt with CPF exposure was $38 \pm 17\%$, while pre-adapted females survived at a rate of $59 \pm 6.6\%$.

It is noted that this value was higher than that from female groups at 15 ppt (both, pre-adapted and non-adapted), where survival rates were between $47 \pm 11\%$ and 43 ± 15 respectively.

5.3.3) Comparison of male and female survival receiving the same treatments

In general, male harpacticoids showed a significantly higher sensitivity towards the various stresses (Figure 5.3, student-test), with no survival noted during the combined stress of CPF exposure and salinity decrease. Without CPF addition, unadapted male harpacticoids were also significantly more sensitive towards salinity decrease, with a survival rate of only $46 \pm 8\%$, compared to the survival rate of $81 \pm 8\%$ of unadapted female harpacticoids.

5.3.4) Comparison with field measurements

The measured salinities in the Lourens River estuary (Figure 5.4a) showed seasonal fluctuations with higher salinities during the summer season as well as considerable fluctuations in-between the sampling intervals. The salinity values within the estuary ranged from 0 ppt during winter season (August) up to 38 ppt during summer season (February).

Measured CPF concentration associated with suspended sediments showed fluctuations from 0 µg/kg up to 38.54 µg/kg (Figure 5.4b) with highest concentrations during spring/summer seasons (October, February, March).

CPF concentrations higher than the average measured value applied in the present study (5.69 µg/kg) were measured during 37% of the sampling dates.

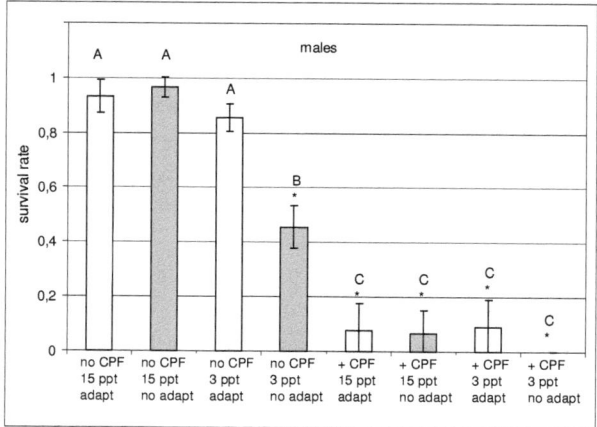

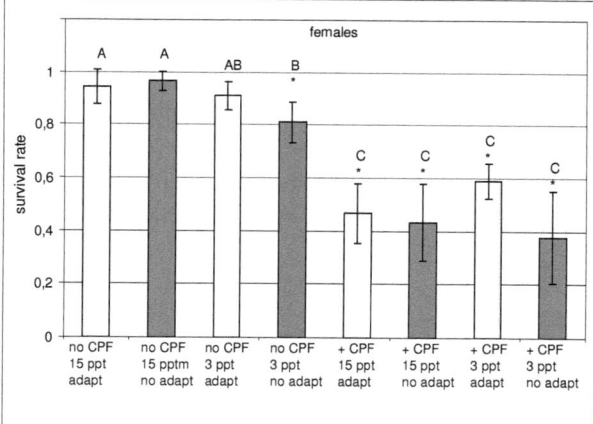

Figure 5.3. Survival rates of male and female *Mesochra parva* in relation to the experimental treatments. Different letters indicate significant differences due to various combinations of experimental variables (Tukey's HSD test, $P < 0.05$), asterisks denote significant differences in survival of male and female organisms after the same treatment (Student's *t*-test, $P < 0.05$). Each column represents the mean of six replicates +/- standard deviation.

Interactive effects

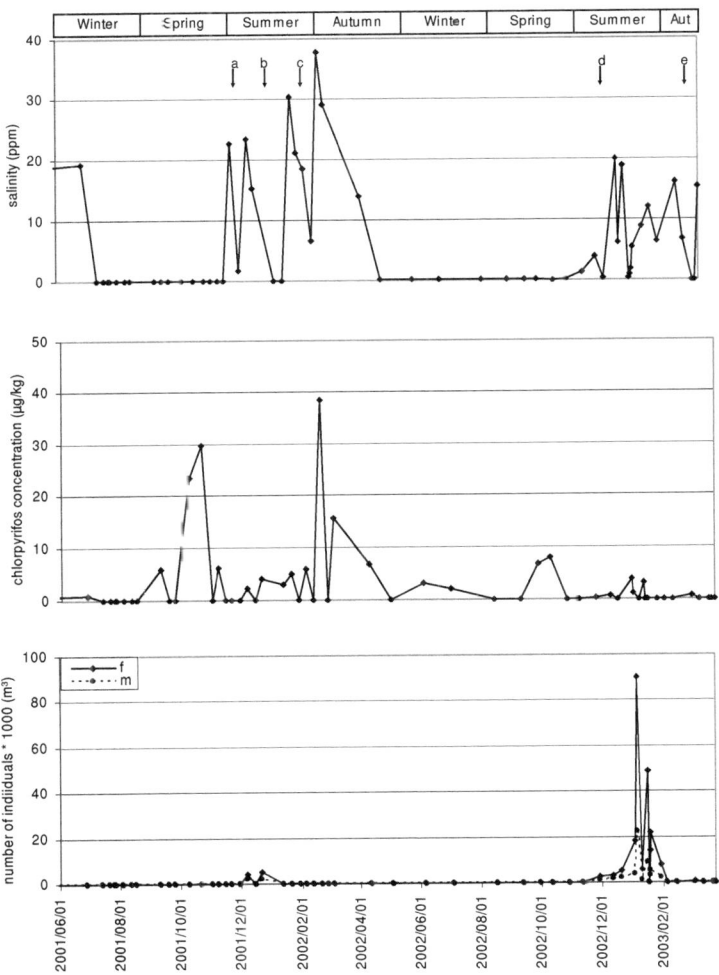

Figure 5.4. Measurements of salinity at the bottom (a), particle associated chlorpyrifos concentration (b) and *Mesochra parva* abundance (c) measured in the lower reaches of the Lourens River estuary between October 2001 and March 2003. Arrows in (a) indicate runoff events with amount of rainfall (a = 3 mm, b = 46.2 mm, c = 2.6 mm, d = 5.2, e = 51.8 mm).

The abundance of male organisms was always lower than that of females (Figure 5.4c), which correlates with the higher sensitivity of males during the laboratory studies. No *M.*

parva specimens (male or female) were found during the sampling dates with CPF concentrations higher than 5.69 µg/kg. However, no direct correlation was found between salinity, CPF exposure and abundance of *M. parva* in the field, due to the high variability of the system (Bollmohr, unpublished data).

5.4 DISCUSSION

5.4.1) Water quality and CPF analysis

All measured physico-chemical water parameters would be expected to have no deleterious effects on the test species, since no significant difference between the different treatments was recorded.

It is well known that, because of higher ionic strength, high salinities can "crystallise" hydrophobic organic chemicals from the water to the sediment phase. Therefore salinity is the driving factor for the bioavailability of pesticides in marine and brackish environments (Chapman and Wang, 2001). Our study supports this finding, since higher salinity resulted in higher CPF concentration in the sediment and lower concentration in the water compared to lower salinity. This was probably due to enhanced desorption of pesticide from the sediment at the lower salinity. This observation is of field relevance, considering that a runoff event, which transports particle-associated CPF into the estuary, also decreases salinity (as indicated by Figure 5.4a) and therefore might increase the bioavailability of CPF.

5.4.2) Copepod survival in relation to experimental factors

In the groups without CPF exposure, reduced salinity in combination with adaptation has a significant influence on the survival of males, but not females. Similar findings on lower male survival at hypoosmotic conditions were reported for other copepod species by Chen *et al.* (2006), in *Pseudodiaptomus annandalei* and by Staton *et al.* (2002) in *Microarthridion litorale*. In the present study, *M. parva* males that had not been pre-adapted to daily salinity fluctuations, survived at a lower rate (46 ± 7.8%) at 3 ppt than of those that had previously been pre-adapted (86 ± 5.0%). In females, the survival rate of the non-adapted species (81 ± 7.8%) is also lower compared to the pre-adapted ones (91 ± 5.4%), but the difference was statistically not significant ($p = 0.15$).

Our results demonstrate that animals pre-adapted to fluctuating salinity had an advantage in terms of survival during salinity change during a runoff event in the Lourens River estuary. However it needs to be clarified that the salinity change from 15ppt to 3ppt is very site specific and the advantage of pre-adapted animals needs to be verified for other salinites. This means that *M. parva* necessarily possesses suitable adaptation mechanisms to adjust the ion- and osmoregulation in varying conditions. Generally we may distinguish two ways of acclimatory responses (Willmer *et al.* 2000): mechanisms allowing an immediate reaction, e.g., in unstable environments, and mechanisms that enable a euryhaline organism to perform optimally over an anticipated range in ambient salinity. The latter is usually connected to a period of adaptation.

Unfortunately, nothing is known about the particular osmoregulatory mechanisms in *M. parva*. However, after 20 days of acclimatisation to fluctuating salinities another harpacticoid copepod, *Tigriopus californicus*, had accumulated high amounts of free amino acids, namely proline and alanine (Goolish and Burton, 1988). Specimens not experiencing such abrupt daily changes contained a factor of 5-6 lower proline and alanine levels. The authors concluded that such elevated levels of free amino acids would make osmoregulation more independent from fluctuations in inorganic ions, particularly Na^+ and K^+, minimizing their perturbing effect on enzyme regulation. Such a capability of proline and alanine accumulation and retention in response to fluctuating salinities would require some period of adaptation. Mechanisms involved in this adaptation process include elevated biosynthesis of enzymes involved in the metabolism of proline and alanine and adjustments of membrane permeability by rearrangements of lipid structure and transporter proteins (of ions, water, amino acids, etc) (Goolish and Burton, 1988, Pequeux, 1995, Willmer *et al.*, 2000).

The lower survival rate of non-adapted male copepods at 3 ppt salinity in the present study suggests that immediate response mechanisms are less effective than those that pre-adapted animals were able to develop during their acclimatization period. Non-adapted males probably had to rely solely on energy-consuming processes to keep ionic- and osmotic balance, particularly with respect to the major cations Na^+ and K^+, after rapid salinity decrease. The most important transporter system for Na^+ and K^+ is the membrane bound Na^+/K^+-ATPase system, which keeps internal cellular sodium levels low and potassium levels high (Pequeux, 1995). It may also be involved in Na^+ uptake into the haemolymph from dilute environments but its operation is energetically costly in terms of ATP-consumption. Also, further ATP-consuming processes may be involved in ionoregulation (Pequeux, 1995).

The assumption that non-adapted male *M. parva* had no alternate or additional options (or only at much lesser efficiency) to keep ionic- and osmotic balance during the present experimental situation (rapid salinity decrease with or without CPF exposure), and thus had to rely on energy-consuming processes, coincides well with our results. Non-adapted females, on the other hand, seem to be more tolerant to abrupt salinity decrease, which may be attributable to their generally higher energy resources (Klosterhaus *et al.*, 2003).

Following CPF exposure, survival of *M. parva* was reduced to values between 0% and 15% in males, and to 38% to 60% in females (Figure 5.3). As a single experimental factor, CPF at measured levels between 5.0 and 7.0 µg/kg sediment resulted in the highest decrease in survival rate of male and female *M. parva* copepods. A microcosm toxicity test on intact estuarine sediment with the marine copepod *Amphiascus tenuiremis* (Chandler *et al.*, 1997) showed acute effects after 96 h at CPF levels ranging from as high as 21 to 33 µg/kg, and a study by Chandler and Green (2001) found that chronic full life-cycle exposures to concentrations of 11-22 µg/kg sediment-associated CPF resulted in consistently significant reductions in *A. tenuiremis* copepodit and naupliar production..

Our results clearly demonstrated that adaptation to cycling salinities was beneficial for male survival, and probably also for females. However, our hypothesis that adaptation to fluctuating salinities would also lead to enhanced survival when exposed to a combination of CPF and hypoosmotic stress was not supported by the results. No significant differences related to adaptation and low salinity were found among the CPF contaminated groups (Figure 5.3). However, the fact that none of the non-adapted males at a salinity of 3 ppt survived CPF exposure might suggest that this group was nevertheless more sensitive to the stressor combination than the pre-adapted male group with CPF addition, where survival was 8.9%. Simultaneously, non-adapted females survived CPF exposure at a salinity of 3 ppt at a rate of 38%, while pre-adapted female survival was 59%. Hall and Anderson (1995) stated that the effect of the combination between salinity change and toxicity of organic chemicals is species-specific and Sprague (1985) argued that this effect is dependent on the strategies that euryhaline organisms use to maintain osmotic balance. Oligochaetes that cannot regulate the internal fluid volume, when exposed to fluoranthene (a polycyclic aromatic hydrocarbon) exhibited higher mortalities at higher salinities (Weinstein, 2003). In contrast, the grass shrimp *P. pugio*, an osmoregulator maintaining nearly constant haemolymph sodium levels across a wide range of salinities, did not show any variation in mortality rates across the salinities tested (Knowlton and Schoer, 1984). Staton *et al.* (2002) reported, however, a clear interaction between salinity and CPF exposure when testing the sensitivity of the meiobenthic harpacticoid copepod *Microarthridion littorale*, without including salinity adaptation as a third factor. In general marine species tend to be more sensitive to pesticide compounds than freshwater species (Wheeler *et al.*, 2002).

CPF sorption to the sediment was different at the two salinity levels and led to unequal rates of CPF-partitioning between water and sediment (Table 5.2). Particularly in the water phase after 96 h CPF concentrations differed by more than one third (0.9 μg/L at 15 ppt and 1.5 μg/L at 3 ppt salinity). One could assume that pre-acapted species gain advantage from the lower CPF concentration in the sediment during lower salinity, since salinity did not pose stress, which could mean that sediment is the main CPF exposure source for these particle-feeding organisms.

Remarkably, the variances among the replicates from the CPF-exposed groups (males and females) were higher than amongst the replicates without CPF addition. The reason for this, if any, is not known but as a consequence these higher variances within the data lowered the discriminating power of a comparison of group means by ANOVA.

5.4.3) Comparison of male and female survival rates

This study demonstrated a higher sensitivity of unadapted male *M. parva* towards abruptly decreased salinity and to CPF exposure. Higher sensitivity of male copepods compared to females to contaminant exposure has been documented previously (Klosterhaus *et al.*, 2003; D Pinto *et al.*, 1993). Also, the study by Staton *et al.* (2002) resulted in a higher sensitivity of males towards CPF exposure concentrations than females. Such results can be explained by females either accumulating contaminants slower than males, eliminating contaminants from their tissues faster, or by possessing higher lipid contents (Lassiter and Hallam, 1990, Staton *et al.*, 2002; Klosterhaus *et al.*, 2003). Higher amounts of storage lipids would (i) allow a better (or longer lasting) energy supply for energy-consuming osmoregulatory processes, and (ii) allow them to sequester more ingested CPF into lipid pools, where it would be kept away from sensitive target enzymes, particularly acetylcholine esterase (Klosterhaus et al., 2003; Lassiter and Hallam 1990). Furthermore, females are generally larger than males, resulting in lower surface area-to-volume ratios. Therefore they are likely to sequester less contaminants than males (Raisuddin et al., 2007). These considerations, however, have to remain speculative as no details on the osmoregulation and energy supply mechanisms in *M. parva* are available. It should be noted that a female-biased sex ratio can also be explained as an adaptive response to high disturbance conditions in which density-dependent mortality is of lesser importance than density–independent mortality. Sustek (1984) indicated that the sex ratio of a given arthropod population will probably change as a consequence of increasing anthropogenic pressure, since the relative proportion of males and females is an important population-adaptive parameter related to the intrinsic rate of population growth. He suggested that under conditions of high disturbances (i.e. density-independent mortality) the populations with the capability of adjusting their sex ratios in the sense of increasing their intrinsic rate of population growth would presumably be favoured. This could also explain the substantial higher abundance of female M. parva in the Lourens River estuary (Figure 5.4c).

5.4.4) Comparison with field measurements

The fact that pre-adapted harpcaticoids were less affected by salinity decrease, with females exhibiting a higher tolerance, coincides with the fact that female-dominated populations of this euryhaline species were found during field samplings at a wide range of salinities between 1 ppt and 38 ppt (Figure 5.4c). This study showed that CPF exposure at measured concentrations of 5.89 to 6.38 µg/kg was comparable to field conditions in the Lourens estuary, South Africa (Figure 5.4b) (Bollmohr et al., 2007), and had a clear effect on the survival rate of males and females. As detected by field analyses, CPF concentrations in the Lourens River estuary exceeded the concentration used in the laboratory experiments at 37% of all sampling dates (Figure 5.4b). No *M. parva* copepods were found at these times in the field (Figure 5.4c). However, laboratory and field results are not always easy to compare. In microcosm experiments using whole sediment samples derived from a pristine estuarine system, Chandler et al. (1997) found no effects on meiofauna populations spiked at the same concentrations between 21 to 33 µg/kg CPF, though these concentrations had been effective in laboratory studies. They concluded that associated organic matter, microbial community and other faunas may decrease bioavailability and persistence of contaminants in the microcosms, compared with single-species bioassays.

5.5 CONCLUSION

Results indicated that a CPF exposure of 5.89 – 6.38 µg/kg, which was exceeded in 37% of the sampling in the field, posed a detrimental effect on the survival rate of *M. parva* in the laboratory. Additionally there was an indication that adaptation to fluctuating salinities was beneficial for both males and females. On the other hand, the hypothesis that adaptation to fluctuating salinities would also lead to enhanced survival when exposed to a combination of CPF and hypoosomotic stress was not supported. However a trend was observed suggesting a higher sensitivity of non pre-adapted organisms to CPF and hypoosomotic stress, compared to unadapted organisms, which requires further elucidation. This study suggests that (at least the males of) *M. parva* copepods, living at a constant high salinity, will be more seriously impacted by a runoff event (CPF exposure x salinity decrease) than organisms experiencing daily salinity fluctuations. This observation has important implications for the management of temporarily open estuaries in South Africa, regarding regulation of freshwater abstraction from rivers and water storage reservoirs.

CHAPTER 6

COMPARISON OF ENVIRONMENTAL VARIABLES, ANTHROPOGENIC STRESSORS AND MEIOFAUNA COMMUNITY STRUCTURE IN TWO TEMPORARILY OPEN ESTUARIES DIFFERING IN SIZE AND TYPE OF CATCHMENT

S. Bollmohr, P.J. van den Brink, P.W. Wade, J.A. Day, R. Schulz

Submitted to Water SA

6.1 INTRODUCTION

Temporarily open/closed estuaries (TOCE) are the dominant estuary type in South Africa, comprising 73% of all estuaries (Whitfield, 1992). Similar systems, sometimes referred to as "blind", 'intermittently open", or "seasonally open" estuaries", are also found in Australia, on the west coast of the USA, South America and India (Ranasinghe and Pattiaratchi, 1999). TOCE's are temporally closed off to the sea, and the lack of freshwater input may result in a number of changes in chemical and physical variables (e.g. salinity, temperature, flow) and anthropogenic stressors (e.g. pesticides). This spatial and temporal variability may have an impact on biotic variables (e.g. zooplankton, meiofauna, ichthyofauna, phytoplankton) (Kibirige and Perissinotto, 2003). Hardly any long-term monitoring data exist for these estuaries making it difficult to understand the variability of the estuaries and factors influencing the estuarine communities. A few studies sampled different variables on a monthly basis (Nozais et al., 2005, Kibirige and and Perissinotto, 2003) and some only on a seasonal basis (Perissinotto et al., 2000) over a period of up to a year. These studies indicate high temporal variability within these systems. This study is one of the first studies looking at natural variables, anthropogenic stressors and community response over a period of two years. One of the limited factors of understanding South African estuaries is the lack of long term monitoring data. The National Water Act (Act 36 of 1998) in South Africa recognises basic human water requirements as well as the need to sustain the country's freshwater and estuarine ecosystems in a healthy condition for present as well as future generations (Adams et. al., 2002). National Monitoring Programmes for freshwater systems are in place since many years (DWAF, 2006) but no programme is designed yet for estuaries.

Many estuarine and coastal management initiatives worldwide, e.g. in North America, Europe and Australia, are required to derive and use environmental quality indices, e.g. implementation of the European Water Framework Directive and the UC Clean Water Act (Quintino et al., 2006). Those indices are then required to test for departure from a reference or control situation. Thus to be able to identify impacts of anthropogenic stressors like pesticides, it is important to understand the natural spatial and temporal variability in physical and chemical conditions and its impact on community structure. Therefore it is crucial to understand the driving variables of communities in natural estuarine systems (reference sites), the catchments of which are not associated with any anthropogenic activity. To achieve this, the dynamic of abiotic and biotic variables of a natural system was compared with the dynamic of variables of a disturbed system using a multivariate approach namely Principal Response Curve. The Principal Response Curve (PRC) approach has been shown to be valuable in statistically analyzing multiple endpoints (van den Brink and ter Braak, 1999). The statistical methodology has been used already for biomonitoring by other authors (Den Besten & van den Brink, 2005; van den Brink *et al.*, 2009).

The study area in this investigation is the Lourens River estuary (Western Cape, South Africa) representing a disturbed system and the Rooiels River estuary representing the reference system. Previous studies (Bollmohr and Schulz, 2009; Schulz *et al.*, 2001) have shown that pesticides are contributing most to the pollution and associated effects in the Lourens River. Particles entering the estuary were shown to be associated with high concentrations of chlorpyrifos (19.6 μg/kg), endosulfan (18.6 μg/kg) and prothiofos (34.0 μg/kg) (Bollmohr *et al.*, 2007), with highest concentrations during the spring and summer season.

This study focused on community changes of organisms living in the flocculent layer between sediment and water column, including epi-benthic, hyper-benthic and demersal meiobenthos and zooplankton organisms. The importance of these organisms in anthropogenic stressor studies is discussed in Bollmohr *et al.*, 2009c. Internationally many studies on meiofauna community structure have focused on the natural distribution and its driving variables like sediment composition (Sherman and Coull, 1980), salinity (Miliou, 1993) and food source (e.g. Decho and Castenholz, 1986). To determine the effect of pesticides on meiobenthic abundances various studies were performed using sediment toxicity tests (e.g. Chandler and Green, 2001; Bollmohr *et al.*, 2009b) or microcosms (Chandler *et al.*, 1997) but only a few studies have used biomonitoring in the field (Warwick *et al.*, 1990, Bollmohr *et al.*, 2009c). Few studies have so far focused on seasonal fluctuations of meiofauna and zooplankton in South African estuaries in relation to the distribution of physical factors (Nozais *et al.*, 2005, Pillay and Perissinotto, 2008), state of the estuaries mouth (Kibirige & Perissinotto, 2003) and food web interactions (Perissinotto *et al.*, 2000), and hardly any study exist in relation to pollutants (Bollmohr *et al.*, 2007; Bollmohr *et al.*, 2009c).

The aim of this study was
- to identify the driving environmental variables (including natural and anthropogenic stressors) in a reference system versus a disturbed system
- to identify if and how these variables change the community structure of the flocculent layer.

6.2 MATERIALS AND METHODS

6.2.1) Study area

The study took place in the Western Cape, South Africa (Figure 6.1). This area is characterised by a high rainfall in winter and a dry summer, as is typical of the Mediterranean climate of the region. The two temporarily open estuaries chosen for the study enter into False Bay, a nearly 40km wide bay southeast of Cape Town. The estuaries differ mainly in the size and kind of catchments:

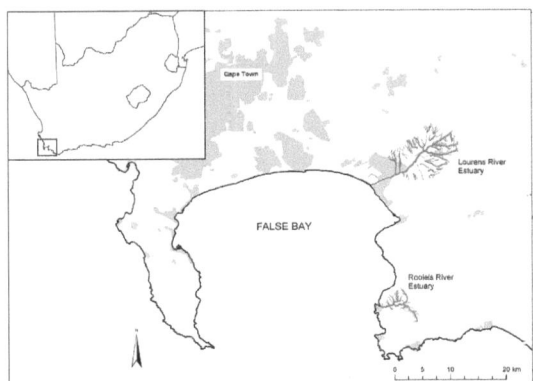

Figure 6.1 False Bay and the location of Lourens River and Rooiels River estuary

Lourens estuary

The Lourens River mouth is located at 34°06'S and 18°49'E in the northeast corner of False Bay. The river (20 km long) rises in the Hottentots Holland Mountains, flows through intensive agricultural areas and through the town of Somerset West, after which it enters False Bay at the town of Strand (Cliff and Grindley, 1982). The estuary is about 0.710 km^2 in size with a tidal range of 1.48 m. Mean Annual Runoff is approximately 122 x 10^6 m^3 (Whitfield and Bate, 2007). Land-use in the Lourens River sub-catchment of 140 km^2 consists of forestry, agriculture, residential areas and light industries. A large section of the upper catchment is privately owned agricultural land with vineyards and apple, pear and plum orchards on which pesticide application takes place between August and mid February before fruit harvest (Schulz, 2001).

Rooiels estuary

The Rooiels River mouth is situated at 34°18'S and 18°49'E. The Rooiels River catchment lies within the southern extension of the Hottentots Holland Mountain reserve and flows through the Kogelberg Nature Reserve. The total distance from the river mouth to the end of the longest tributary is only 9 km. (Whitfield and Bate, 2007). The estuary is about 0.122 km² in size with a tidal range of 1.48 m. The Mean Annual Runoff is 10 x 10^6 m^3. The entire Rooiels catchment (21 km²), lies within a nature reserve and has small holiday residential areas but no associated agricultural activities (Heydorn and Grindley, 1982).

6.2.2) Pesticide application and seasonal weather data

The 20 month investigation period (August 2001 till March 2003) comprised all four seasons with dry summer season with infrequent rainfalls (Figure 6.2) from November to December and wet winter seasons from June to August. Organophosphates, such as chlorpyrifos (686 kg/ha) and prothiofos (87 kg/ha), organochlorines such as endosulfan (158 kg/ha) and pyrethroids such as cypermethrin (8 kg/ha) and fenvalerate (5 kg/ha) are frequently applied to pears, plums and apples between August and February before fruit harvest (Dabrowski et al., 2002) with a high frequency of application during the dry summer season from November to January (Schulz, 2001) (indicated by the dotted lines in Figure 6.2).

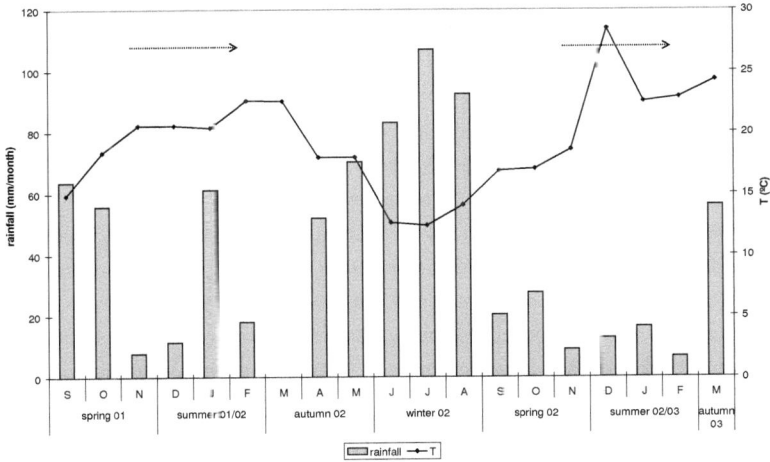

Figure 6.2. Monthly rainfall and temperature data (obtained from Verlegen weather station at 80m altitude) during the study period from September 2001 to March 2003. (The dotted lines indicated the spraying season of pesticides).

6.2.3) Physico-chemical measurements

Various physicochemica parameters were measured every two weeks during the study period. The measurements took always place in the morning during low tide. The flow was measured every meter across the estuary using a flow meter (Hoentzsch Co, Waiblingen, Germany) and the outflow was calculated.

Dissolved oxygen, pH, temperature and conductivity were measured with electronic meters from Wissenschaftliche Technische Werkstaetten GmbH, Weilheim, Germany. The salinity was obtained by converting the conductivity values into salinity using an automatic converter (Fofonoff and Millard, 1983). Total Suspended Solids (TSS) was measured with a turbidity meter (Dr. Lange, Duesseldorf, Germany) and nutrients were detected by photometric test kits from Macherey & Nagel, Dueren, Germany. The total organic carbon (TOC) was measured using a quantitative method which is based upon the removal of all organic matter followed by gravimetric determination of sample weight loss (ASTM, 2000). Grain size analysis was performed by dry fraction sieving in samples previously treated with hydrogen peroxide. Sediment were divided into seven fractions: silt and clay (< 0.063 mm), very fine sand (0.063-0.125 mm), fine sand (0.125-0.25 mm), medium sand (0.25-0.5 mm), coarse sand (0.5-1 mm), very coarse sand (1-2 mm) and gravel (>2 mm). Each fraction was weighed and expressed as percentage of total weight. Four sediment samples (top 15 cm) per site (within a 1 m² transect) were taken and analysed.

Microalgal biomass was determined in the water and in the sediment by measuring the concentration of their photosynthetic pigment, chlorophyll a. For phytoplankton analysis in the water, four replicates of 2 L water samples were filtered through plastic Millipore filter towers with Whatman (GF/C) glass-fibre filters. The chlorophyll a was extracted by mortar the filters with 10 ml of 99% acetone. After centrifuging the light absorbance at 665nm of the supernatant was determined, before and after adding two drops of 0.1 N HCL, using a UV-VIS spectrophotometer. Chlorophyll a was calculated using the equation of Hilmer (1990). Nine replicates of benthic microalgal samples were taken within a 1 m² transect with a 15 ml syringe. The chlorophyll a was extracted from the top 1.5 cm of sediment in the same way as the chlorophyll a in the water.

6.2.4) *Pesticide analysis*

Samples of insecticides (every two to three week) associated with suspended particles were accumulated continuously by a suspended-particle sampler, from which they were collected every two to three weeks. The suspended-particle sampler consisted of a plastic container (500 mL) with a screw-on lid containing a hole (2 cm in diameter), which contained an open glass jar that was stored directly under the hole in the lid. The samplers were stored approximately 5 cm above the river bed by being attached to a metal stake which was fixed into the sediment.

Suspended particles samples were extracted twice with methanol and concentrated using C18 columns. Insecticides were eluted with hexane and dichlormethane and were analysed at the Forensic Chemistry Laboratory, Department of National Health, Cape Town. Measurements were made with gas chromatographs (Hewlett-Packard 5890, Avondale, PA, USA) fitted with standard Hewlett-Packard electron-capture, nitrogen phosphorus and flame-photometric detectors, with a quantification limit of $0.1 \mu g\ kg^{-1}$ and overall mean recoveries between 79 and 106% (Schulz *et al.*, 2001). Further details of extraction and analysis of pesticides is described in Schulz *et al.* (2001).

6.2.5) Meiobenthic sampling and community analysis

A different sampling approach was applied to reduce the sampling error caused by small-scale variability in sediment (Livingston, 1987) and to sample the organisms living in the flocculent layer between sediment and water column (including hyper-, epi-benthic, and demersal organisms). A stretch of 10 m flocculent layer (width: 30 cm) was sampled actively just above the bottom sediment using a zooplankton net of 63 µm mesh size. Four replicates were taken and within each replicate the zooplankton and meiobenthos were identified to the lowest taxonomic level possible. Zooplankton and meiobenthos abundance was determined based on eight sub-samples. Sub-samples were derived using a matrix box with 64 chambers, of which 8 chambers were randomly selected for identification and counting of all organisms.

To assess temporal and spatial patterns of benthic faunal assemblages the following community metrics were calculated: density (Ind/m^3), no of taxa, H' diversity (Shannon & Wiener, 1948).

6.2.6) Statistics

Multivariate techniques are often used to link field concentrations of chemicals and other environmental parameters with invertebrate responses (Leps and Smilauer, 2003). In this paper ordination methods are used for dimension reduction, i.e. to reduce the data set to a two-dimensional summary (ordination diagram). In this way a graphical summary of the data set is obtained. Redundancy Analysis (RDA) is the direct form of PCA that enables the researcher to focus the analysis on that particular part of the variance that is explained by external explanatory variables. In our example we used environmental variable explaining a significant part of the variation in abundance of the invertebrates as explanatory variables. The significance of the environmental variables was evaluated using Monte Carlo permutation tests (Van den Brink et al., 2003). Redundancy Analysis was performed separately for the two estuaries.

The differences in community composition between the Lourens and Rooiels river estuaries were visualised by PRC (Principal Response Curves; Van den Brink and Ter Braak, 1999) using the CANOCO software package version 4.5. The PRC is based on the Redundancy Analysis ordination technique (RDA), the constrained form of Principal Component Analysis. The analysis results in a diagram showing time on the x-axis and the first Principal Component of the differences in community structure on the y-axis. This yields a diagram showing the deviations in time between the two sites, with the

Rooiels estuary as reference. The species weights are shown in a separate diagram, and indicate the affinity the species have with this dominant indicated difference. The species with a high positive weight are indicated to show a response similar to the response indicated by PRC, those with a negative weight, one that is opposite to the response indicated by PRC. Species with a near zero weight are indicated to show a response very dissimilar to the response indicated by PRC or no response at all.

The data sets of the physico-chemical variables and pesticides were tested for significant difference using the student t-test and were also analysed using PRC. In these analyses the variables were centred and standardised before analysis to account for differences in measurement scale (Kersting and Van den Brink, 1997). How PRC can be used to analyse monitoring data is described more in detail in Van den Brink *et al.* (2009).

6.3 RESULTS

6.3.1) Temporal and spatial comparison of environmental variables and particle associated pesticides

Due to the different type and size of catchment, many environmental variables in the Lourens River estuary were significantly different from the variables in the Rooiels River estuary, as shown in Table 6.1. The most significant differences ($p<0.001$) were calculated for depth, salinity, temperature, nitrate, pH and TSS. In general, the Lourens River estuary is characterised by a significantly lower depth (0.51 ± 0.18 m compared to 1.07 ± 0.14 m) salinity at the bottom (6.29 ± 8.18 ppt compared to 19.5 ± 11.6 ppt) and the surface (1.69 ± 2.07 ppt compared to 5.19 ± 5.74 ppt) indicating a larger associated catchment with a higher amount of runoff. However, bottom temperature (19.4 ± 4.43 ºC) and surface temperature (19.2 ± 4.44 ºC) in the Lourens River estuary showed higher values than in the Rooiels estuary (18.0 ± 4.20 ºC and 16.8 ± 3.30 ºC respectively). Total suspended solids are significantly higher in the Lourens River estuary (63.9 ± 39.6 mg/L compared to 1.03 ± 1.21 mg/L), due to higher erosion. Due to less natural associated catchment characterised by fynbos vegetation and associated humic acids, the pH is significantly higher in the Lourens River estuary than in the Rooiels River estuary.

Table 6.1 Mean (± standard error) physicochemical parameters measured in the Rooiels River and Lourens River estuary from September 2001 to March 2003 (n=44) (Stars indicating the significant difference between the two estuaries (* $p < 0.05$; ** $p < 0.01$; *** $p < 0.001$))

	Unit	Lourens River estuary	Rooiels River estuary
Outflow	m^3/sec	1.00 ± 1.15	0.46 ± 0.19
Mean depth ***	m	0.51 ± 0.18	1.07 ± 0.14
Salinity (surface)***	ppt	1.69 ± 2.07	5.19 ± 5.74
Salinity (bottom)***	ppt	6.29 ± 8.18	19.5 ± 11.6
Temperature (surface)***	°C	19.2 ± 4.44	18.0 ± 4.20
Temperature (bottom)***	°C	19.4 ± 4.43	16.8 ± 3.30
Nitrate***	mg/l	3.47 ± 3.36	0.00 ± 0.00
Nitrite*	mg/l	0.077 ± 0.124	0.003 ± 0.006
Ammonia	mg/l	0.007 ± 0.013	0.01 ± 0.02
Ortho-phosphate*	mg/l	0.39 ± 0.32	0.17 ± 0.25
Oxygen (surface)	mg/l	9.33 ± 1.88	8.75 ± 1.28
Oxygen (bottom)*	mg/l	8.80 ± 2.00	7.75 ± 1.01
pH***		7.58 ± 0.51	6.71 ± 0.98
Total Suspended Solids***	mg/l	63.9 ± 39.6	1.03 ± 1.21
Total organic carbon	%	1.29 ± 0.62	1.29 ± 0.54
Grain size**	μm	164 ± 79.9	239 ± 30.1
Chlorophyll a (water)**	mg/l	3.76 ± 3.98	0.39 ± 0.46
Chlorophyll a (sediment)**	mg/l	8.98 ± 8.45	2.29 ± 1.75

As a result of anthropogenic inputs, nitrate, nitrite and orthophosphate concentration are significantly higher at the Lourens River estuary (p<0.001; p<0.01; p<0.01). Phytoplankton biomass measured as chlorophyll a concentrations in the water (3.76 ± 3.98 mg/l compared to 0.39 ± 0.46 mg/l) and in the sediment (8.98 ± 8.45 mg/m^2 compared to 2.29 ± 1.75 mg/m^2) is significantly higher as well (p<0.01; p<0.01).

Cypermethrin, fenvalerate, endosulfan total (END), p,p-DDE, chlo pyrifos (CPF) and prothiofos were found in varying concentrations during almost all months within the Lourens River estuary (Table 6.2, Figure 6.3) with highest concentrations during autumn (S,O,N) and spring (M, A, M). No pesticide concentrations were expected in the Rooiels River estuary due to no associated agricultural catchment. However, CPF, PTF and cypermethrin concentrations were detected frequently with highest concentrations during the summer months (N, D, J) with even higher prothiofos concentrations than in the Lourens River estuary (Table 6.2, Figure 6.3). Therefore only fenvalerate, total - END, p,p-DDE and azinphos-methyl are significantly different between the two estuaries.

Table 6.2 Mean (± standard error) particle bound pesticide concentrations (μg/kg) in the Lourens and Rooiels River estuary and 90th percentile from September to March 2003 (n=27). (Asterisk indicating the significant difference between the two estuaries (* $p < 0.05$; ** $p < 0.01$))

	Lourens River estuary			Rooiels River estuary		
	mean	% of detection	90th percentile	Mean	% of detection	90th percentile
Cypermethrin	0.92 ± 1.09	92	2.84	0.31 ± 0.84	23	0.42
Fenvalerate*	0.19 ± 0.22	77	0.64	0.00 ± 0.00	0	0.00
DDE**	8.42 ± 13.6	5	31.0	0.00 ± 0.00	0	0.00
Endosulfan total**	6.03 ± 9.40	86	18.6	0.00 ± 0.00	0	0.00
Azinphos-methyl	0.07 ± 0.31	50	0.00	0.00 ± 0.00	0	0.00
Chlorpyrifos	7.04 ± 10.5	92	19.6	3.21 ± 5.46	50	8.90
Prothiofos	6.74 ± 15.8	85	34.0	12.90 ± 28.03	39	22.0

Prothiofos (34 μg/kg) showed the highest concentrations in the Lourens River followed by p,p-DDE (31.0 μg/kg), CPF (19.6 μg/kg), total - END (18.6 μg/kg), cypermethrin (2.84 μg/kg) and fenvalerate (0.64 μg/kg) (Table 6.2). However cypermethrin (92%) and CPF (92%) showed the highest frequency of detection, followed by total - END (86%) and prothiofos (85%).

The Rooiels River estuary showed highest concentration in prothiofos (22.0 μg/kg), followed by CPF (8.9 μg/kg) and cypermethrin (0.42 μg/kg). Chlorpyrifos showed the highest frequency of detection (50%), followed by prothiofos (39%) and cypermethrin (23%).

Natural versus disturbed estuary

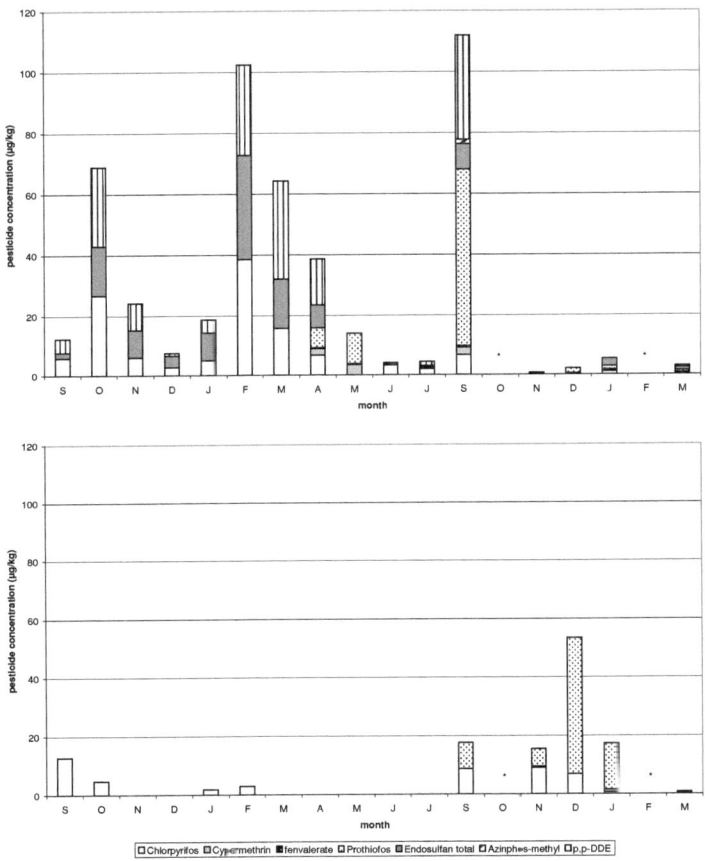

Figure 6.3. Monthly composition of particle bound pesticide concentrations detected at Lourens and Rooiels River estuary from September 2001 to March 2003 Asterisks indicate that no samples were taken.

The PRC diagram (Figure 6.4) indicates differences in environmental variables and particle bound pesticide concentrations between the two estuaries and their importance in contributing to the difference. There is a high temporal variability, since 42% of the variation is attributed to the difference between sampling dates. However, spatial difference explained 58% of the variation. Variables contributing mostly to the difference were higher concentrations of endosulfan, p,p-DDE and nitrate concentrations in the Lourens River estuary and larger grain size and higher salinity at the bottom in the Rooiels River estuary. Furthermore, flow, nitrite, chlorpyrifos and chlorophyll are important variables contributing to the difference between the two estuaries.

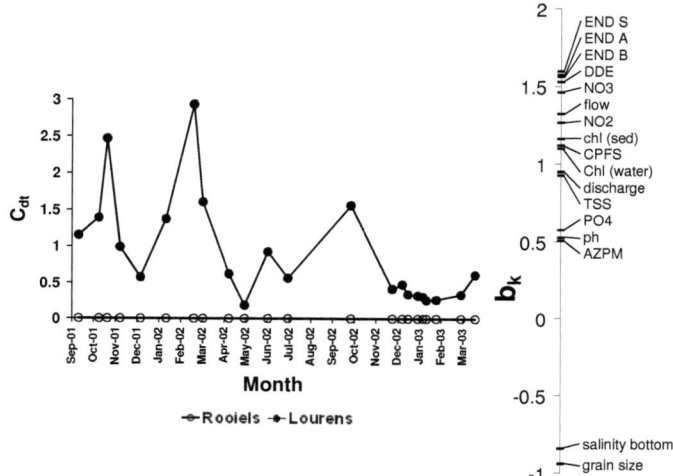

Figure 6.4. Diagram showing the first component of the PRC of the differences in measured environmental variables and pesticides between Lourens and Rooiels River estuary. Forty-two percent of the total variation could be attributed to differences between sampling dates, the other 58% to differences between the estuaries; 43% of the latter is displayed in the diagram. The parameter weights shown in the right part of the diagram represent the affinity of each parameter with the response shown in the diagram. For clarity, only parameters with a weight > 0.5 or < -0.5 are shown.

6.3.2) Temporal and spatial comparison of meiofauna communities

In general the meiofauna community in the Rooiels River estuary showed a significantly higher number of taxa ($p < 0.001$), a significant higher Shannon Wiener Diversity Index ($p<0.001$) and a generally lower abundance with less variability than in the Lourens River estuary (Table 6.3). The PRC (Figure 6.5) showed that the two community structures were very different from each other, of which 52% was explained by temporal differences and 48% by differences between the two estuaries. The differences were mostly explained by a higher abundance of Cypretta and Darcythompsonia in the Rooiels River estuary and a higher abundance of Thermocyclops and Canthocamptus in the Lourens River estuary. Generally the difference is attributed to the abundance of more species in the Rooiels River estuary, which correlates with the higher diversity index (Table 6.3).

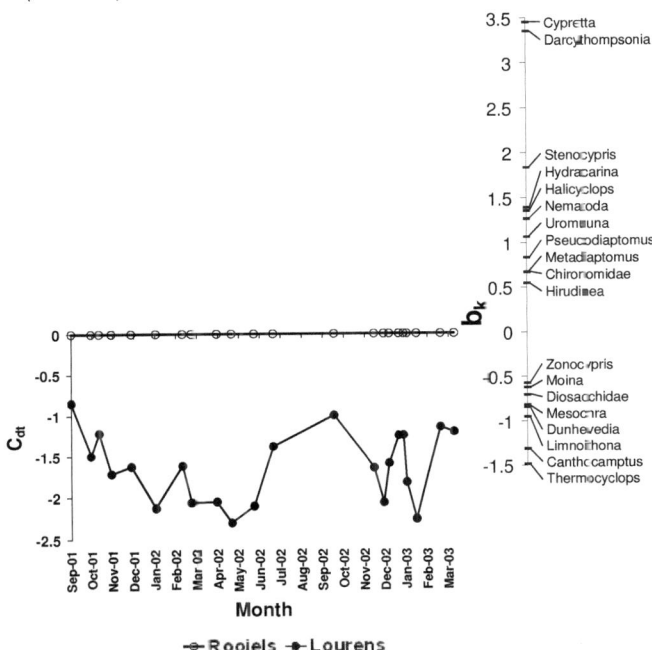

Figure 6.5. Diagram showing the first component of the PRC of the differences in species composition between the Rooiels and Lourens rivers. Fifty-two percent of the total variation in species composition could be attributed to differences between sampling dates, the other 48% to differences in species composition between the estuaries; 44% of the latter is displayed in the diagram. The taxa weights shown in the right part of the diagram represent the affinity of each taxa with the response shown in the diagram. For the sake of clarity, only species with a weight larger than 0.5 or smaller than -0.5 are shown.

Table 6.3 Mean (± standard error) number of taxa, total abundance and Shannon Wiener Diversity Index in the Lourens and Rooiels River estuary and 90th percentile from September to March 2003 (n=27). (Asterisks*** indicating the significant difference between the two estuaries $p < 0.001$))

	Lourens River estuary		Rooiels River estuary	
	mean	90th percentile	Mean	90th percentile
Number of taxa***	6.8 ± 2.7	10	10.3 ± 2.7	14
Total abundance	6600 ± 12391	16423	2442 ± 1318	3951
Shannon Diversity Index***	1.1 ± 0.4	1.6	1.6 ± 0.3	1.9

6.3.3) Meiofauna community and their driving variables

Within the RDA biplot the most important taxa and significant environmental variable are shown (Figure 6.6 and 6.7). The variables explaining a significantly part (14%) of the seasonal variation in invertebrate abundance in the Rooiels River estuary were salinity and temperature (Figure 6.6). The RDA indicated that most of the taxa were shifted towards high salinity and temperatures. Taxa like *Upogebia*, *Nereis*, *Uroma* and nematodes were clearly indicated to be positively correlated to salinity and temperature. No taxa clearly decreased with increasing salinity (oligohaline) and temperature. Thus in general this estuary was dominated by estuarine (euryhaline) and marine taxa.

The meiofauna community in the Lourens River estuary is significantly influenced by more variables due to more anthropogenic impacts (Figure 6.7). The variables explaining a significant part of the variance in the dataset (43%) were salinity and temperature on the one hand and chlorpyrifos, nitrate and flow on the other hand. *Mesochra* decreased with higher chlorpyrifos and nitrate concentrations, and to increase with higher salinity and temperature. Furthermore the taxa *Pseudodiaptomus*, *Canthocamptus* and polychaetes were negatively correlated with CPF and nitrate concentrations. In general, more taxa decreasing with higher salinity (like *Moina* or *Limnoithona*) indicating an abundance of more oligohaline taxa.

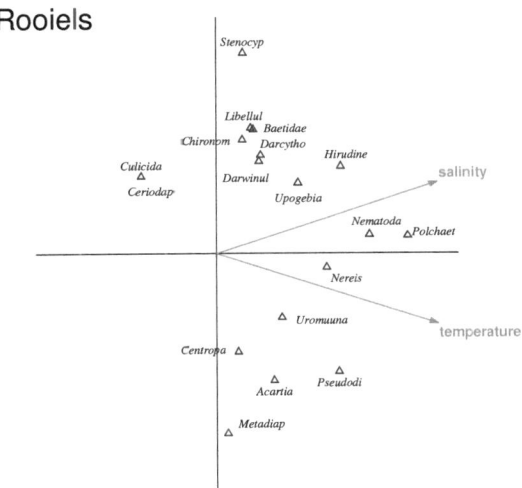

Figure 6.6. RDA triplot resulting from an analysis using the environmental variables and pesticide concentrations that explain a significant part of the seasonal variation in the species composition in the Rooiels River estuary as explanatory variables and meiofauna taxa composition as taxa data. The two significant variables explain 14% of the seasonal variation in species composition, of which 71% is displayed on the first axis and another 29% on the second one. For clarity only the 18 most important taxa out of 33 are shown.

Natural versus disturbed estuary

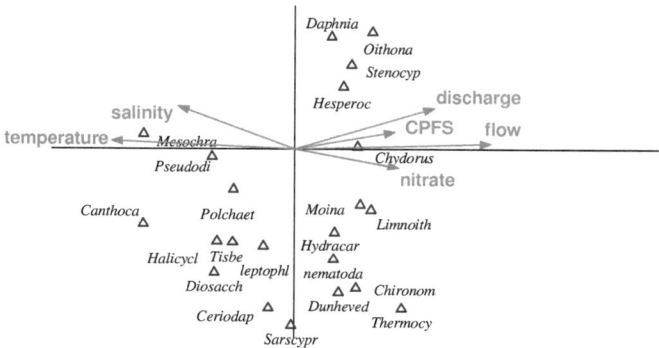

Figure 6.7. RDA triplot resulting from an analysis using the environmental variables and pesticide concentrations that explain a significant part of the seasonal variation in the species composition in the Lourens River estuary as explanatory variables and meiofauna taxa composition as taxa data. The six significant variables explain 43% of the seasonal variation in species composition, of which 69% is displayed on the first axis and another 12% on the second one. For clarity only the 20 most important taxa out of 33 are shown.

6.4 DISCUSSION

6.4.1) *Temporal and spatial comparison of environmental variables and particle associated pesticides*

Due to its catchment having no associated agricultural and urbanization influences, the Rooiels estuary is more pristine than the Lourens River estuary, with lower temperature, TSS, pH and particle bound pesticide concentrations, and also with lower nutrient concentrations resulting in lower chlorophyll a concentrations as a measure for food availability. The lower Mean Annual Runoff results in a more saline environment with evident stratification in salinity, temperature and oxygen with higher salinity values and lower oxygen values at the bottom of the system, which is typical for natural deep TOCE's during the dry phase. Kibirige and Perissinotto (2003) detected a salinity ranging from 0 to 26 ppt with higher values at the bottom. Furthermore the Rooiels River estuary is much deeper than the Lourens River estuary, which could enhance stratification (Whitfield and Bate, 2007).

However, although no agricultural catchment was associated, pesticides like CPF, prothiofos and cypermethrin were found in the system, particularly during the summer season of 2003, with concentration higher than those at the Lourens River estuary. The input of pesticides can have various explanations which require further elucidation. (1) Studies have shown that volatilization can cause off-site transport of pesticides via atmospheric transport and subsequent contamination of surface waters (Harman-Fetcho et al., 2000; LeNoir et al., 1999). The Western Cape climate is characterised by high wind speeds during summer in a south-easterly wind direction (towards Rooiels River estuary) and most application via mist blow takes place during the summer season (Schulz 2001), which coincides with the high concentrations found during summer. However, high mountains (Hottentots Holland Mountain) are situated between the two catchments which could serve as a barrier for atmospheric pesticide transport. Additionally it is questionable why endosulfan was not found in the Rooiels River estuary, since it is also heavily applied during the summer season. (2) Due to the south-easterly wind direction, the current within False Bay is directed from the Lourens River estuary (and other estuaries with associated agricultural catchments) towards the Rooiels River estuary (Taljaard et al., 2000). Particle bound pesticides with a high Koc and low solubility (as prothiofos, CPF and cypermethrin) could be transported via the current within the False Bay and could enter the Rooiels River estuary via the tide. However, pesticides are likely to bind on fine particles which are likely to settle down either already within the estuary (Bollmohr et al., 2009c) or within the bay. Additionally it is questionable why p,p-DDE which is present in high concentrations in the Lourens River with the highest K_{oc} and longest DT_{50} (Bollmohr et al., 2007) has not been detected in the Rooiels River estuary (3) Residential areas are situated on the southern side of the estuary and the houses are mainly used during holidays (Heydorn and Grindley, 1982). The summer holidays are during December and January, which coincides with the time at which high pesticide concentrations were found. Many household products, which are heavily used in South Africa contain chlorpyrifos and cypermethrin.

Bollmohr et al. (2007) ranked the pesticide concentrations detected in the Lourens River in terms of their risk towards the environment by comparing with various effect concentrations (Chandler and Green, 2001; Chandler et al., 1994; Leonard et al., 2001). Chlorpyrifos and total - END posed the highest risk towards freshwater and marine communities in the Lourens River estuary. Within the Rooiels River estuary only CPF may pose a risk towards a harpacticoid species (Bollmohr et al., 2009b), although it is far below any effect levels determined by other authors (Chandler and Green, 2001). No sufficient toxicity data and threshold values were found for prothiofos, thus it was not possible to assess a possible risk for this pesticide towards the aquatic environment in both estuaries.

6.4.2) Temporal and spatial comparison of meiofauna communities

The prolonged period of mouth closure that TOCE's experience generally leads to poor levels of taxonomic diversity (Perissinotto et al., 2000). Hartwell & Clafin (2005) classified a Shannon Wiener Index of below 1.5 for macrobenthos in permanently open estuaries as "low" and above 2.4 as "high". However, a Shannon Wiener index for zooplankton in an impacted temporarily open estuary ranged from 1.00 to only 1.20 (Kibirige and Perissinotto, 2003). Therefore both estuaries still showed a higher index. However, although the Lourens River estuary received more freshwater species (Bollmohr et al., 2009c) due to the higher Mean Annual Runoff, the diversity is still lower than in the Rooiels River estuary which is characterised by a higher number of taxa and a lower abundance. The lower standard deviation (and therefore lower variability) in abundance in the Rooiels River estuary might be a result of less disturbance (Warwick and Clarke, 1993).

Species contributing mostly to the spatial difference are *Cypretta* (ostracod) and *Darcythompsonia* (harpacticoid) for the Rooiels estuary (with the highest positive weight) and *Thermocyclops* and *Canthocamptus* for the Lourens River estuary (with the highest negative weight). Ostracods are mainly influenced by temperature, salinity, depth, grain size (Curry, 1999) and sewage pollution as a result of low oxygen (Mezquita et al., 1999). They are not significantly more sensitive to pollution than other crustaceans (Sanchez-Bayo, 2006) based on toxicity data but they have a longer recovery rate (Dernie et al., 2003). However, the combination of low oxygen, salinity, low grain size (mud), higher temperature and more frequent pesticide concentration might have resulted in the loss of ostracods at the Lourens River estuary. No information is available on the harpacticoid genus *Darcythompsonia* and further work on the sensitivity of this genus is required. *Thermocyclops* and *Canthocamptus* are known as freshwater and/or estuarine species and might have been carried into the Lourens River estuary from upstream, whereas the salinity in the Rooiels River estuary might have been too high to survive. No sufficient toxicity data are available for these genera in order to reach conclusions of the causality of the correlations of their abundances with the pesticide levels in both estuaries.

6.4.3) Meiofauna community and their driving variables

The RDA performed separately for the two estuaries identified salinity and temperature as the only significant variables driving the community structure in the flocculent layer for the Rooiels River estuary. These variables are already identified as the most important variables for natural estuarine systems by various authors (Austen, 1989; Willmer et al., 2000; Nozais et al., 2005). Grain size and water depth are often regarded as important variables to determine meiobenthic organisms (Soetaert et al., 1994; Curry, 1999; Sherman and Coull, 1980). However, these variables hardly showed any variation in the Rooiels River estuary.

The community structure of the Lourens River estuary is, however, significantly influenced by a number of variables, including salinity, temperature, flow and anthropogenic stressors like nitrate and CPF, which may suggest interactions between effects of natural variables and anthropogenic variables. The abundance of the genus Mesochra as an example increased with decreasing CPF concentrations and increasing salinity. A microcosm toxicity test on intact estuarine sediment with the marine copepod *Amphiascus tenuiremis* (Chandler et al., 1997) showed acute effects after 96 h at CPF levels ranging from as high as 21 to 33 µg/kg. Also a study by Chandler and Green (2001) found that chronic full life-cycle exposures to concentrations of 11-22 µg/kg sediment-associated CPF resulted in consistently significant reductions in *A. tenuiremis* copepodit and naupliar production. These concentrations were exceeded by the 90th percentile found in the Lourens River estuary (19.6 µg/kg). Another study by Bollmohr et al. (2009b) indicated that a CPF exposure of 5.89 – 6.38 µg/kg, which was exceeded in 37% of the sampling in the field in this study, posed a detrimental effect on the survival rate of *M. parva* in the laboratory. During this study the survival decreased by 60 -90%

due to chlorpyrifos exposure and increased due to higher salinity (from 3 - 15ppt) by 10-50%.

One may suggest that the frequency and mean of CPF in the Rooiels River estuary (50%; 3.21 ± 5.46) did not influence the community structure, whereas the frequency and mean concentration of CPF concentrations in the Lourens River estuary (92%; 7.04 ± 10.5 µg/kg) did. This can be a result of interactive effects with natural variables, other pollutants or simply the higher concentration and frequency of detection in the Lourens River.

6.5 CONCLUSION

This study successfully confirmed the results of Bollmohr *et al.*, 2007, suggesting CPF and total - END posing the highest risk towards an estuarine system. We also confirmed the interactive effect of salinity and chlorpyrifos exposure and the importance of temperature and salinity in natural estuarine system by using RDA. Although the Rooiels River estuary showed some concentrations of particle bound pesticides, none of them influence the community structure in flocculent layer significantly. However the source of pesticides required more elucidation.

CHAPTER 7

GENERAL DISCUSSION

General Discussion

Agricultural impact on water quality in the coastal zone through sedimentation (Valette-Silver *et al.*, 1986) and eutrophication (Carpenter *et al.*, 1998) have been well documented internationally over the past half-century. Exposure and effect assessments of pesticides, however, have generally focused on freshwater systems and do not provide a holistic understanding of the behavior of pesticides along an entire river stretch, of which estuaries are the endpoints. These systems, particularly temporarily open estuaries, experiencing specific dynamics (mainly in salinity), which might influence the occurrence and effect of pesticides (Chapman and Wang, 2001).

Particle-bound pesticides may pose a major environmental hazard, primarily because they may enter the estuary frequently during runoff events, they may accumulate in estuaries and they may be more available to estuarine organisms than to freshwater organisms, due to different feeding behavior (suspension and deposit feeding). However, hardly any studies have been performed in order to determine the hazard of particle-bound pesticides.

Central to this thesis is the question of what concentrations of particle-bound pesticides do enter the estuary, how do they behave in interaction with other natural variables and which pesticides pose the highest risk towards benthic communities. Chapter 3 – 6 have each focused on integral components of a multiple-evidence based risk assessment of particle-bound pesticides: Sediment contamination in comparison with guideline values including EqP modelling (Chapter 3), field investigations of sediment toxicity and benthic community structure (Chapter 4,6) and laboratory spiked sediment toxicity tests (Chapter 5). This was done in an attempt to understand the i) spatial and temporal variability of natural variables, particle-bound pesticides and the response of the meiobenthos, as well as ii) the interactive effect of some natural variables and pesticides towards the benthic community. During these studies salinity was one of the variables that reflected high importance together with exposure and effect of particle-bound pesticides.

7.1 Exposure Assessment

Three chapters discussed the spatial and temporal variability of particle associated pesticide concentrations. Pesticides like chlorpyrifos (32.2 µg/kg), endosulfan (191 µg/kg) and p,p-DDE (131 µg/kg) were found in higher concentration than in other studies (Bergamaschi et al., 2001, Doong et al., 2002) whereas fenvalerate (1.78 µg/kg) was found in lower concentrations (Chandler et al., 1994). Prothiofos (34 µg/kg) and cypermethrin (12.34 µg/kg) were detected the first time in an estuarine system. A distinct temporal (Chapter 3) and spatial (Chapter 4) variability in pesticide concentrations was established.

Highest concentrations were found during spring season (Chapter 3) which coincides with peak application time of pesticides in the Western Cape. This supports the finding of Steen et al. (2001), who established a link between pesticides detected in the Scheldt estuary and the time of application. Furthermore during this season inflow decreases and as a result salinity starts to increase, thus higher pesticide concentration might be linked to higher salinity. Sujatha et al., (1999) explained a temporal variability of pesticides in an Indian estuary by changes in salinity. Bondarenko et al. (2004) determined a longer persistence of chlorpyrifos caused by higher salinity due to inhibition of microbial degradation. Furthermore, it is well known that because of higher ionic strength, high salinities can "crystallise" hydrophobic organic chemicals from the water to the sediment phase (Chapman & Wang, 2001), which supports the higher chlorpyrifos concentrations in the sediment during elevated salinities in the sediment toxicity test (Chapter 5). This is of high field relevance, considering that a runoff event, which transports particle-bound pesticides into the estuary, also decreases salinity and thus might decrease the pesticide absorbed to particles, and increase the concentration in the water. One of the variables influencing the quantity of pesticide that can be expected in surface runoff is the time interval between the application of pesticides and the first heavy rainfall event (Cole et al., 1997). This was supported by the high input of pesticide after the first runoff in the spraying season, compared to no pesticide input after the second runoff event (Chapter 4).

Within the multivariate statistics the temporal variability often explained the difference between sites along the estuary with higher significance than the spatial variability. However there was a distinct difference along the Lourens River estuary with mostly higher concentration within the upper reaches, indicating a reduction in pesticide concentrations (Chapter 4). Two reason could explain this spatial decrease: i) The high amount of macrophytes could have sorbed the pesticides, as it has been shown for azinphos-methyl (Dabrowski et al., 2006) or chlorpyrifos (Moore et al., 2002); ii) Less

General Discussion

contaminated marine particles could have diluted the contaminated riverine particles further downstream, as it has been observed by Turner and Millward (2002).

Chapter 3 compared the particle-bound pesticide and calculated dissolved pesticide concentrations using the Equilibrium Partitioning Theory (DiToro, 1991) with existent sediment and water quality guidelines. The relevance of the equilibrium partitioning theory for organisms is still contended. The EqP theory assumes that a chemical bound to sediment organic carbon is in thermodynamic equilibrium with the chemical dissolved in the aqueous phase of the sediment (interstitial water) and the lipid components of the exposed organism (McCauley *et al.*, 2000). Some authors argue that the approach underestimates the bioavailable fraction because it does not include the molecules bound to particles, which can be taken up through ingestion by deposit feeding benthic organisms (Selck *et al.*, 2003.).This is particularly the case in marine/ estuarine systems with a high percentage of deposit-feeding organisms frequently ingesting sediment particles (Weston *et al.*, 2000). Other authors argue that the approach overestimates bioavailability, because it exceeds the freely dissolved concentration that drives the uptake into an organism (Batley and Maher, 2001). However the lack of sediment toxicity data urges legislators to use the EqP-method to estimate sediment quality guideline values from aquatic toxicity data (van Beelen *et al.*, 2003).

Although the guideline values derived by applying the EqP approach in this study were less conservative than the guideline values derived by the Canadian (CCME, 1999) and Australian Water Quality Guidelines (ANZECC, ARMACANZ, 2000), they were mostly exceeded by the calculated endosulfan and chlorpyrifos concentrations in the Lourens River estuary, indicating a high potential risk towards the organisms. Only particle-bound endosulfan (0.43 – 12.7 µg/kg), however, exceeded the sediment quality guidelines set by US EPA (2002) and ANZECC, ARMACANZ (2000), indicating a higher protection level for chlorpyrifos by the water quality guidelines.

Although the Rooiels estuary was chosen as a reference site due to no associated agricultural activities, prothiofos, chlorpyrifos and cypermethrin has been detected mainly during the summer season. This supports the suggestion by Whitfield and Bate (2007) that, although the condition of a Reference site should refer to natural

General Discussion

unimpacted ecological state, it is important to realise that in some estuaries changes have occurred.

Prothiofos is one of the pesticides which was detected frequently in elevated concentrations in both estuaries. However it was not possible to determine its risk towards estuarine organisms, since no sufficient toxicity data are available in the general literature. This organophosphate is only registered in Australia, New Zealand and South Africa and not in Europe and US.

7.2 Effect Assessment

Natural variables

Salinity is one of the most important variables impacting the benthic community in both studied estuaries (Chapter 6). Diversity indices are generally lower in temporarily open estuaries, especially during the closed status of the estuary with no exchange with the seawater and lack of marine species (Kibirige & Perissinotto, 2003). However, higher salinity resulted in an increase in mostly estuarine taxa, which supports the findings of previous studies (Austen, 1989; Perissinotto *et al.*, 2002, Viliou, 1993). Salinity is one of the main environmental factors controlling species distribution, abundance and sex-ratio in marine, and particularly in estuarine organisms (Willmer *et al.*, 2000).

The results of Chapter 5 demonstrate that the test organism *M. parva* (mainly males) pre-adapted to fluctuating salinity had an advantage in terms of survival during salinity decrease, highlighting the importance of regular salinity change for estuarine organisms. Similar findings on lower male survival at hypoosmotic conditions were reported for other copepod species by Cher *et al.* (2006), in *Pseudodiaptomus annandalei* and by Staton *et al.* (2002) in *Microarthridion litorale*. The authors concluded that elevated levels of free amino acids namely proline and alanine (Goolish and Burton, 1988) would make osmoregulation more independent from fluctuations in inorganic ions, particularly Na^+ and K^+, minimizing their perturbing effect on enzyme regulation. Such a capability of proline and alanine accumulation and retention in response to fluctuating salinities would require some period of adaptation. Non-adapted females, on the other hand, seem to be more tolerant to abrupt salinity decrease, which may be attributable to their generally higher energy resources (Klosterhaus et al., 2003).

TOC also drove the spatial difference along the Lourens River estuary (Chapter 4), supporting findings by Sherman and Coull (1980). However, temperature and grain size, which are often stated as important variables contributing to seasonal variability in estuarine communities (Nozais et al., 2005) showed a low importance in the Lourens River estuary but a high importance in the Rooiels River estuary (Chapter 6).

Particle bound pesticides

The importance of considering the interaction of natural and anthropogenic variables is highlighted by the following examples. Chapter 3 indicates a higher sensitivity of marine taxa than freshwater taxa towards the pesticides detected in the Lourens River estuary. During summer season, the mouth is almost closed, with no exchange with the seawater and lack of marine species (Kibirige & Persissinotto, 2003). This would suggest a less severe impact during the summer season with a lower occurrence of marine species. However, estuarine benthic species are affected by higher salinity during pesticide exposure, which is also supported by the fact that the survival rate of exposed adapted *M. parva* decreased with higher salinity. Pesticide concentrations bound to particles increase during higher salinity due to "crystallization" from the water to the sediment phase (Chapman & Wang, 2001) and due to longer persistence (Bondarenko et al. 2004), thus particle-feeding organisms (like most benthic organisms) ingesting these sediment particles are more exposed to pesticides during the spring/ summer season when the salinity is increasing. This is also shown by the opposite effect of the significant variables salinity and chlorpyrifos driving the benthic community and correlates with the finding by Hall and Anderson (1995) stating that the toxicity of organophosphate insecticides appeared to increase with salinity.

The bioavailability within the organism remains unclear, however. The bioavailability of the pesticide may not explain the interaction between exposure and salinity increase. Hall and Anderson (1995) stated that the effect of the combination between salinity change and toxicity of organic chemicals is species-specific and Sprague (1985) argued that this effect is dependent on the strategies that euryhaline organisms use to maintain osmotic balance. To conclude, although the benthic estuarine community is positively influenced by higher salinity during the dry season, they might be negatively influenced by the combination of salinity increase and pesticide exposure.

The PRC analysis provides information which is not provided by general community indices, which could be used to determine pesticide effects on meiobenthos communities in estuaries. Temporal variability in the dataset explained more variation in community structure, environmental variables and pesticide concentrations than spatial variability, already stated by Kibirige & Persissinotto (2003). However, the effects of pesticides have been successfully distinguished from the impact of natural variables by the PRC analysis, by comparing two runoff events. An increase in spatial differences in community structure was revealed by PRC for both the first and second runoff events. The spatial difference in community structure during the first runoff event was probably caused by pesticide concentrations and the difference during the second event was caused by environmental variables. The RDA performed separately for the "Reference" and the "impacted" estuary, identified salinity and temperature as the only significant variables driving the community structure in the flocculent layer for the Rooiels River estuary. These variables are already identified as the most important variables for natural estuarine systems by various authors (Austen, 1989; Willmer *et al.*, 2000; Nozais *et al.*, 2005). The community structure of the Lourens River estuary is, however, significantly influenced by a number of variables, including salinity, temperature, flow and anthropogenic stressors like nitrate and CPF, which may suggest interactions between effects of natural variables and anthropogenic variables.

Most of the meiofaunal organisms, however, tend to be r-selected, typically exhibiting small body size, short life cycle, early reproduction, rapid development, variable population size, density-independent mortality, low competitive ability, and high reproduction (Chapman and Wang, 2001). Thus the high potential of recovery, which is also reflected in the high temporal dynamics in the field, needs to be considered. Furthermore, estuarine species adapt to the high variability in the system and become tolerant of changes, including the presence of organic contaminants (Dauvin, 2007), which could promote resistance and tolerance towards contaminants.

General Discussion

Chapter 3, 4 and 6 suggest that the pesticides endosulfan and chlorpyrifos exhibit the greatest impact in the Lourens River estuary. Although chlorpyrifos was detected in the Rooiels River frequently as well, only 17% of all samples exceeded the effect level of 5.7 µg/kg. As a result, chlorpyrifos did not affect the community structure as strongly as in the Lourens River estuary. Furthermore, salinity was much higher in the Rooiels River estuary, which might impact the benthic community much more than infrequent pesticide exposure.

Only a few studies have investigated changes in estuarine community structures due to pesticide exposure. Results of the study done by deLorenzo *et al.* (1999) suggested that exposure to agricultural pesticides can lead to both functional and structural changes in the estuarine microbial food web. Endosulfan primarily reduced bacterial abundance, the number of cyanobacteria and phototrophic biomass. Chlorpyrifos decreased the abundance of protozoan grazers. Chandler & Green (2001) found a decrease in production of *Amphiascus tenuiremis* copepodit and naupliar during CPF exposure of 11-22 µg/kg. A predicted sediment quality criterion for a maximum safe concentration of chlorpyrifos was calculated to be 31.2 µg/kg. Leonard *et al.* (2001) determined a 10-d NOEC of 42 µg/kg for sediment associated endosulfan for the epibenthic mayfly *Jappa kutera*. Chandler & Scott (1991) suggested that polychaete (*Streblospio benedicti*) populations in the field may be strongly depressed by sediment endosulfan concentration of 50 µg/kg. The pesticide concentrations at which effects should be observed in estuarine systems suggested by various authors (Chandler & Scott, 1991; Chandler *et al.*, 1997; Chandler & Green, 2001; Bollmohr and Schulz, 2009) are frequently exceeded in the Lourens River estuary.

7.3 Management implications

Sediment quality guidelines values for chlorpyrifos of 32.2 µg/kg (US EPA, 2002) and 53.0 µg/kg (ANZECC, ARMACANZ, 2000) are far above the acute effect level determined during the toxicity test in this study (5.70 µg/kg). Furthermore it was suggested that chlorpyrifos is one of the pesticide influencing the benthic community most severely, which coincides with the fact that 38% of the samples taken in the Lourens river estuary exceeded the effect level. This would suggest that the sediment quality guidelines might be under-protective in light of the fact that the test organisms *M.*

parva, responding most to the increase in chlorpyrifos concentrations in the field and being used in the sediment toxicity test, plays a central role in marine ecological and physicochemical processes (Hicks and Coull, 1983; Coull, 1990).

The effect of a runoff event was measured in the field and simulated during the toxicity test. It was recognised that during the first runoff event after high application time of pesticides which resulted mostly in an increase in chlorpyrifos and endosulfan concentrations caused the highest deviation in community structure, mainly due to the change in the abundance of M. parva. During the sediment toxicity test a combination of salinity decrease and chlorpyrifos exposure was used in order to simulate a runoff event. During this combination the survival rate of adapted organisms was higher during low salinity than during high salinity. This would suggest that the combination of salinity decrease and chlorpyrifos exposure do not pose a synergistic risk towards estuarine organisms. However, the frequent salinity change remains important for the adaptation ability and their survival potential in the system. This observation has important implications for the management of temporarily open estuaries in South Africa, regarding regulation of freshwater abstraction from rivers and water storage reservoirs (Scharler and Baird, 2003)

Furthermore, one could assume that the pre-adapted organisms gain advantage from the lower chlorpyrifos concentrations during lower salinity, since salinity did not pose any stress. This could suggest that sediment is the main chlorpyrifos exposure source for these particle feeding organisms, which would support the argument that the EqP theory might underestimate the risk towards deposit feeding taxa (Serck et al., 2003).

Data produced in this research thus provide important information to understand the spatial and temporal variability of pesticides and its interaction with natural variables in a temporarily open estuary. To summarise, this study indicated that pesticides like endosulfan and chlorpyrifos posed a risk towards benthic organisms in a temporarily open estuary in particular during spring season. Furthermore an important link between pesticide exposure/ toxicity and salinity was identified.

CHAPTER 8

REFERENCES

LITERATURE CITED IN ALL CHAPTERS

Adams, J.B., Bate, G.C., Harrison, T.D., Huizinga, P., Taljaard, S., van Niekerk, L., Plumstead, E.E., Whitfield, A.K., Wooldridge, T.H. 2002. A method to assess the freshwater inflow requirements of estuaries and application to the Mtata estuary, South Africa. *Estuaries and Coasts* **25**: 1382-1393.

Aldenberg, T.; Jaworska, J.S. 2000. Uncertainty of the hazardous concentration and fraction affected for normal species sensitivity distributions. *Ecotoxicology and Environmental Safety* **46**: 1-18.

Anger, K. 1996. Salinity tolerance of the larvae and first juveniles of a semiterrestial grapsid crab, *Armases miersii* (Rathbun). *Journal of Experimental Marine Biology and Ecology* **202** 205-223.

Ankley, G.T., Call, D.J., Cox, J.S., Kahl, M.D., Hoke, R.A., Kosian, F.A. 1994. Organic carbon partitioning as a basis for predicting the toxicity of chlorpyrifos in sediments. *Environmental Toxicology and Chemistry* **13**: 621-626.

Antonius, G.F., Byers, M.E. 1997. Fate and movement of endosulfan under field conditions. *Environmental Toxicology and Chemistry* **16**: 644-649.

ANZECC, ARMACANZ 2000. Australian and New Zealand guideline for fresh wand marine water quality. Australian and New Zealand Environment and Conservation Council and Agriculture and Resource Management Council of Australia and New Zealand. Canberra, Australia.

ASTM 2000. Standard Test Methods for Moisture, Ash and Organic matter of Peat and other Organic Soils. Method D 2974-00. American Society for Testing and Materials. West Conshohocken, PA.

Austen, M.C. 1989. Factors affecting estuarine meiobenthic assemblage structure: a multifactorial microcosm experiment. *Journal of Experimental Marine Biology and Ecology* **130**: 167-187.

Bailey, H.C., Deanovic, L., Reyes, E., Kimball, T., Larson, K., Cortright, K., Connor, V., Hinton, D.E. 2000. Diazinon and chlorpyrifos in urban waterways in northern California, USA. *Environmental Toxicology and Chemistry* **19**: 82-87.

Batley, G,E., Maher, W.A. 2001. The development and application of ANZECC and ARMACANZ sediment quality guidelines. *Australasian Journal of Ecotoxicology* **7**: 81-92.

Bergamaschi, B.A., Kuivila, K.M., Fram, M.S. 2001. Pesticides associated with suspended sediments entering San Francisco Bay following the first major storm of water year 1996. *Estuaries* **24**: 368-380.

Bollmohr, S., Day, J.A., Schulz, R., 2007. Temporal variability in particle-bound pesticide exposure in a temporarily open estuary, South Africa. *Chemosphere* **68**: 479-488.

Bollmohr, S, Thwala, M, Jooste, S, Havemann, A. 2009a. Report: An Assessment of Agricultural Pesticides in the Upper Olifants River Catchment. Report No. N/0000/REQ0801. Resource Quality Services, Department of Water Affairs and Forestry, Pretoria, South Africa.

Bollmohr, S., Hahn, T., Schulz, R. 2009b. Interactive effect of salinity decrease, salinity adaptation and chlorpyrifos exposure on an estuarine harpacticoid copepod, *Mesochra parva*, in South Africa. *Ecotoxicology and Environmental Safety* (in press).

Bollmohr, S., van den Brink, P.J., Wade., P.W., Day, J.A., Schulz, R. 2009c. Spatial and temporal variability in particle-bound pesticide exposure and their effects on benthic communities structure in a temporarily open estuary. *Estuarine, Coastal and Shelf Science* (in press).

Bollmohr, S., Schulz, R. 2009. Seasonal changes of macroinvertebrate community in a Western Cape River receiving nonpoint-source insecticide pollution. *Environmental Toxicology and Chemistry* (in press).

Bondarenko, S., Gan, J., Haver, D.L., Kabashima, J.N. 2004. Persistence of selected organophosphate and carbamate insecticides in waters from a coastal watershed. *Environmental Toxicology and Chemistry* **23**: 2649-2654.

Bouwman, H., 2004. South Africa and the Stockholm Convention on persistent organic pollutants. *South African Journal of Science* **100**: 323-328

British Crop Protection Council 1994. Pesticide Manual. 10th Ed. British Crop Protection Council, Croydon, England.

Brock, T.C.M., Van Wijingaarden, R.P.A., Van Geest, G.J. 2000. Ecological risk of pesticides in freshwater ecosystems, Part 2 Insecticides, Alterra Rapport 089, Alterra, Green World Research, Wageningen.

Burton, G. A., Jr. 1991. Assessing the toxicity of freshwater sediments. *Environmental Toxicology and Chemistry* **10**: 1585-1627.

Burton, G.A., Batley, G.E., Chapman, P.M., Forbes, V.E., Smith, E.P., Reynoldson, T., Schlekat, C.E., den Besten, P.J., Bailer, A.J., Green, A.S., Dwyer, R.L. 2002. A weight-of-evidence framework for assessing sediment (or other) contamination:

Improving certainty in the decision-making process. *Human and Ecological Risk Assessment* **8**: 1675-1696.

Carpenter, S.R., Varaco, N.F., Correll, D.L., Howarth, R.W., Sharpley, A.N., Smith, V.H., 1998. Nonpoint pollution of surface waters with phosphorus and nitrogen. *Ecological Applications* **8**: 559-568

CCME 1999. Canadian environmental quality guidelines. Canadian Council of Ministers of the Environment. Winnipeg (MB), Canada.

Chandler, T.G., Scott, G.I. 1991. Effects of sediment-bound endosulfan on survival, reproduction and larval settlement of meiobenthic polychaetes and copepods. *Environmental Toxicology Chemistry* **10**: 375-382

Chandler, T.G., Coull, B.C., Davis, J.C. 1994. Sediment- and aqueous-phase fenvalerate effects on meiobenthos: Implications for Sediment Quality Criteria Development. *Marine Environmental Research* **37**: 313-327.

Chandler, G.T., Green A.S. 1996. A 14-day harpacticoid copepod reproduction bioassay for laboratory and field contaminated muddy sediments. In G.K. Ostrander, ed., Techniques in Aquatic Toxicology. Lewis, Boca Raton, FL, USA, pp. 23-29.

Chandler, T.G., Coull, B.C., Schizas, N.V., Donelan, T.L. 1997. A culture-based assessment of the effects of chlorpyrifos on multiple meiobenthic copepods using microcosms of intact estuarine sediments. *Environmental Toxicology and Chemistry* **16**: 2339-2346.

Chandler, T.G., Green, A.S. 2001. Developmental stage-specific life-cycle bioassay for assessment of sediment associated toxicant effects on benthic copepod production. *Environmental Toxicology and Chemistry* **20**: 171-178.

Chapman, P.M. 1996. Presentation and interpretation of sediment quality triad data. *Ecotoxicology* **5**: 327-339.

Chapman, P.M., Wang, F. 2001. Assessing sediment contamination in estuaries. *Environmental Toxicology and Chemistry* **20**: 3-22.

Chen, Q. X. Sheng, J. Q., Lin, Q., Gao, Y. L., Lv, J. Y. 2006. Effect of salinity on reproduction and survival of the copepod *Pseudodiaptomus annandalei* Sewell, 1919. *Aquaculture* **258**: 575-582.

Clark, J.R., Goodman, L.R., Bothwick, P.W., Patrick, J.M., Cripe, G.M., Moody, P.M., Moore, J.C., Lores, E.M. 1989. Toxicity of pyrethroids to marine invertebrates and fish: a literature review and test results with sediment-sorbed chemicals. *Environmental Toxicology and Chemistry* **8**: 393-401.

Clarke, G.M. 1993. Fluctuating asymmetry of invertebrate populations as a biological indicator of environmental quality. *Environmental Pollution* **82**: 207-211.

Cliff, S. & Grindely, J.R. 1982. Report No.17 Lourens (CSW7). In: Heydorn, A.E.F. and Grindley, J.R. (eds). Estuaries of the Cape, part 2, Synopsis of available information on individual systems. *CSIR Report No 416*. Stellenbosch, South Africa: 39pp

Coetzee, J.C., Adams, J. B. & Bate, G.C. 1997. A botanical importance rating of selected Cape estuaries. *Water SA* **23**: 81–93.

Cole, J.T., Baird, J.H., Basta, N.T., Huhnke, R.L., Storm, D.E., Johnson, G.V., Payton, M.E., Smolen, M.D., Martin, D.L., Cole, J.C. 1997. Influence of buffers on pesticide and nutrient runoff from Bermuda grass Turf. *Journal of Environmental Quality* **26**: 1589-1598.

Colloty, B.M., Adams, J.B. and Bate, G.C. 2000. The botanical importance of the estuaries in former Ciskei/Transkei. *WRC Report 812/1/00* 150 pp.

Coull, B.C. 1990. Are members of the meiofaunal food for higher trophic levels? *Transactions of the American Microscopical Society* **109**: 233-246.

Coull, B.C., Chandler, G.T. 1992. Pollution and meiofauna: Field, laboratory, and mesocosm studies. *Oceanography and Marine Biology: An Annual Review* **30**: 191-271.

Curry, B.B. 1999. An environmental tolerance index for ostracods as indicators of physical and chemical factors in aquatic habitats. *Paleogeography, Palaeoclimatology, Palaeoecology* **148**: 51-63

Dabrowski, J.M., Peall, S.K.C., Reinecke, A.J., Liess, M., Schulz, R., 2002. Runoff - related pesticide input into the Lourens River, South Africa: basic data for the exposure assessment and risk mitigation at the catchment scale. *Water Air Soil Pollution* **135:** 265-283.

Dabrowski, J.M., Bennett, E.R., Bollen, A., Schulz, R. 2006. Mitigation of azinphos-methyl in a vegetated stream: Comparison of runoff- and spray-drift. *Chemosphere* **62**: 204-212.

Dall, W., Rothlisberg, P.C., Sharples, D.J., 1990. The Biology of the Penaeidae. *Advances in Marine Biology*. **27**. Academic Press, London, pp. 489.

Dauvin, J.-C. 2007. Paradox of estuarine quality: Benthic indicators and indices, consensus or debate for the future. *Marine Pollution Bulletin* **55**: 271-281.

Decho, A.W. & Castenholz, R.W. 1986. Spatial patterns and feeding of meiobenthic harpactiocid copepods in relation to resident microbial flora. *Hydrobiologia* **131**: 87-96.

DeLorenzo, M.E., Scott, G.I., Ross, P.E. 1999. Effects of the agricultural pesticides atrazine, deethylatrazine, endosulfan, and chlorpyrifos on an estuarine microbial food web. *Environmental Toxicology and Chemistry* **18**: 2824-2835.

Den Besten, P.J., van den Brink, P.J., 2005. Bioassay response and effects on benthos after pilot remediations in the delta of the rivers Rhine and Meuse. *Environmental Pollution* **136**: 197-208.

Department of Water Affairs (DWAF) 1996. South African Water Quality Guidelines (Second Edition), Volume 7: Aquatic Ecosystems, Pretoria.

Department of Water Affairs (DWAF) 2004. Methodology for the Determination of the Ecological Water Requirements for estuaries. Version 2. Final Draft, Pretoria.

Department of Water Affairs and Forestry (DWAF) 2006. National Toxicity Monitoring Programme (NTMP) for Surface Waters Phase 2: Record of Decision. By K. Murray, R.G.M. Heath, J.L. Slabbert, B. Haasbroek, C. Strydom and P.M. Matji. Report No.: N000REQ0505 ISBN No: 0-621-36404-5. Resource Quality Services, Department of Water Affairs and Forestry, Pretoria, South Africa.

Dernie, K.M., Kaiser, M.J., Richardson, E.A., Warwick, R.M. 2003. Recovery of soft sediment communities and habitats following physical disturbance. *Journal of Experimental Marine Biology and Ecology* **285-286**: 415-434.

Dinham B. 1993. The Pesticide Hazard. A global health and environmental audit. Pesticides Trust, Zed press, London

Di Pinto, L.M., Coull, B.C., Chandler, G.T. 1993. Lethal and sublethal effects of the sediment associated PCB Aroclor 1254 on a meiobenthic copepod. *Environmental Toxicology and Chemistry* **12**: 1909-1918.

Di Toro, D.M., Zarba, C.S., Hansen, D.J., Berry, W.J., Swartz, R.C., Cpwan, C.E., Pavlou, S.P., Aller, H.E., Thoma, N.A., Paquin, P.R. 1991. Technical basis for establishing sediment quality criteria for nonionic organic chemicals using equilibrium portioning. *Environmental Toxicology and Chemistry* **10**: 1541-1583.

Domagalski, J.L., Kuivila, K.M. 1993. Distributions of pesticides and organic contaminants between water and suspended sediment, San Francisco Bay, California. *Estuaries* **16**: 416-426.

Doong, R.-A., Peng, C.-K., Sun, Y.-C., Liao, P.L., 2002. Composition and distribution of organochlorine pesticide residues in surface sediments from the Wu-shi River estuary, Taiwan. *Marine Pollution Bulletin* **45**: 246-253.

Dye, A.H., 1983. Composition and seasonal fluctuations of meiofauna in a southern African mangrove estuary. *Marine Biology* **73**:165-170.

European Commission 2002. Technical guidance document on risk assessment in support of the Commission Directive 93/67/EEC on risk assessment for new notified substances and the Commission Regulation (EC) 1488/94 on risk assessment for existing substances and Directive 98/8/EC for the European Parliament and the Council concerning the placing of biocidal products on the market. Brussels, Belgium.

Farmer, D., Hill, I.R., Maund, S.J. 1995. A Comparison of the fate and effects of two pyrethroid insecticides (lambda-cyhalothrin and cypermethrin) in pond mesocosms. *Ecotoxicology* **4**: 219-244.

Fleeger, J.W., Carman, K.R., Nisbet, R.M. 2003. Indirect effects of contaminants in aquatic ecosystems. *The Science of the Total Environment* **317**: 207-233.

Flury, M. 1996. Experimental evidence of transport of pesticides through field soils – a review. *Journal of Environmental Quality* **25**: 25-45.

Fofonoff, T.L., Millard, Jr R.C. 1983. Algorithms for computation of fundamental properties of seawater. Unesco Technical Papers in Marine Science **44**: 53.

Forbes, T.L. and Calow, P. 2002. Species sensitivity distributions revisited: A critical appraisal. *Human and Ecological Risk Assessment* **8**: 473-492.

Gee, M.J., Warwick, R.M. 1985. Effects of organic enrichment on meiofaunal abundance and community structure in sublittoral soft sediments. *Journal of Experimental Marine Biology and Ecology* **91**: 247-262.

Giddings, J.M., Solomon, K.R., Maund, S.J. 2001. Probabilistic Risk Assessment of cotton pyrethroids: II. Aquatic mesocosm and field studies. *Environmental Toxicology and Chemistry* **20**: 660-668.

Gillis, C.A., Bonnevie, N.L., Su, S.H., Ducey, S.L., Huntley, S.L., Wenning, R.J. 1995. DDT, DDD, and DDE contamination of sediment in the Newark Bay Estuary New Jersey. *Archive of Environmental Contamination and Toxicology* **28**: 85-92.

Goolish, E.M., Burton R.S. 1988. Exposure to fluctuating salinities enhances free amino acid accumulation in *Tigriopus californicus* (Copepoda). *Journal of Comparative Physiology* **158**: 99-105.

References

Grathwohl, P. 1990. Influence of organic matter from soils and sediments from various origin on the sorption of some chlorinated aliphatic hydrocarbons: Implications on Koc correlations. *Environmental Science and Technology* **24**: 1687-1693.

Green, A.S., Chandler, T.G., Piegorsch, W.W. 1996. Life-stage-specific toxicity of sediment-associated chlorpyrifos to a marine, infaunal copepod. *Environmental Toxicology and Chemistry* **15**: 1182-1188.

Hall, L.W., Anderson, R.D. 1995. The influence of salinity on the toxicity of various classes of chemicals to aquatic biota. *Critical Reviews in Toxicology* **25**: 281.

Harman-Fetcho, J.A., McConell, L.L., Rice, C.P., Baker, J.E. 2000. Wet deposition and air-water gas exchange of currently used pesticides to a subestuary of the Chesapeake Bay *Environmental Science and Technology* **34**:1462-1468.

Harrison, T.D., Cooper, J.A.G. & Ramm, A.E.L. 2000. State of South African estuaries: geomorphology, ichthyofauna, water quality and aesthetics. State of the Environment Series, Report No. 2. Department of Environmental Affairs and Tourism. 127pp.

Harrison, T.D., Hohls, D.R., Neara, T.P., Webster, M.S. 2001. South African Estuaries: Catchment Land-cover. CSIR Report ENV/S 97128, Stellenbosch South Africa.

Hartwell S.I., Clafin, L.W. 2005. Cluster ananlysis of contaminated sediment data: Nodal analysis. *Environmental Toxicology and Chemistry* **24**: 1816-1834.

Heath RG. and Claassen, M. 1999. An overview of the pesticide and metal levels present in populations of the larger indigenous fish species of selected South African Rivers. *Water Research Commission Report No. 428/1/99*.

Heydorn, A.E.F. & Grindley, J.R. 1982. Estuaries of the Cape, Part I.II. Synopses of available information on individual systems. *CSIR Research Report*

Hicks, G.R.F., Coull, B.C. 1983. The ecology of marine meiobenthic harpactiocid copepods. *Oceanography and Marine Biology: An Annual Review* **21**: 67-175.

Hillmer, T. 1990. Factors influencing the estimation of primary production in small estuaries. PhD thesis, University of Port Elizabeth, South Africa.

Hose, G.C., Hyne, R.V., Lim, R.P. 2003. Toxicity of endosulfan to *Atalophlebia* spp. (Ephemeroptera) in the laboratory, mesocosm, and field. *Environmental Toxicology and Chemistry* **22**: 3062-3068.

Hunt, J.W., Anderson, B.S., Phillips, B.M., Nicely, P.N., Tjeerdema, R.S., Puckett, H.M., Stephenson, M., Worcester, K., de Vlaming, V. 2003. Ambient toxicity due to chlorpyrifos and diazinon in a central Californian coastal watershed. *Environmental Monitoring and Assessment* **82**: 83-112.

Ingersoll, C. G., Ankley, G.T. 1995. Toxicity and bioaccumulation of sediment-associated contaminants using freshwater invertebrates – A review of methods and applications. *Environmental Toxicology and Chemistry* **14**: 1885-1894.

Kennedy, A.D., Jacoby, C.A. 1999. Biological indicators of marine environmental health: meiofauna- a neglected benthic component? *Environmental Monitoring and Assessment* **54**: 47-68.

Kersting, K., van den Brink, P.J. 1997. Effects of the insecticide Dursban4E (active ingredient chlorpyrifos) in outdoor experimental ditches: responses of ecosystem metabolism. *Environmental Toxicology and Chemistry* **16**: 251-259.

Kibirige, I., Perissinotto, R. 2003. The zooplankton community of the Mpenjati Estuary, a South African temporarily open/closed system. *Estuarine, Coastal and Shelf Science* **58**: 727-741.

Klosterhaus, S.L., DiPinto, L.M., Chandler, G.T. 2003. A comparative assessment of azinphos-methyl bioaccumulation and toxicity in two estuarine meiobenthic harpacticoid copepods. *Environmental Toxicology and Chemistry* **22**: 2960-2968.

Knowlton, R.E., Schoen, R.H. 1984. Salinity tolerance and sodium balance in the prawn *Palaemonetes pugio* Hulthuis, in relation to other *Palaemonetes* spp. *Comparative Biochemistry and Physiology* **77A**: 425-430.

Kuivila, K.M., Foe, C.G. 1995. Concentrations, transport and biological effects of dormant spray pesticides in the San Francisco estuary, California. *Environmental Toxicology and Chemistry* **14**: 1141-1150.

Laskowski, D. 2002. Physical and chemical properties of pyrethroids. *Reviews Environmental Contaminant Toxicology* **174**: 49-170.

Lassiter, R.R., Hallam, T.G 1990. Survival of the fattest: implications for acute effects of lipophilic chemicals on aquatic populations. *Environmental Toxicology and Chemistry* **9**: 585-595.

LeNoir, J.S., McConnell, L.L., Fellers, G.M., Cahill, T.M., Seiber, J.N. 1999. Summertime transport of current-use pesticides from California's central valley to the Sierra Nevada Mountain Range, USA. *Environmental Toxicology and Chemistry* **18**: 2715-2722.

Leonard, A.W., Hyne, F.V., Lim, R.P., Leigh, K.A., Le, J., Beckett, R. 2001. Fate and toxicity of endosulfan in Naomi River Water and Bottom Sediment. *Journal of Environmental Quality* **30**: 750-759.

Leps, J., Smilauer P. 2003. Multivariate analysis of ecological data using Canoco. Cambridge University Press, 269 p.

Liess, M., Schulz, R., Neumann, M. 1996. A method for monitoring pesticides bound to suspended particles in a small stream. *Chemosphere* **32**: 1963-1969.

Livingston, R.J. 1987. Field Sampling in Estuaries: The Relationship of Scale to Variability. *Estuaries* **10**: 194-207.

Loague, K., Corwin, D.L. and Ellsworth, T.R. 1998. The challenge of Predicting Nonpoint-source Pollution. *Environmental Science and Technology* **32**: 130-133.

London, L. and Myers, J. 1995. General patters of agrichemical usage in the southern region of South Africa. *South African Journal of Science* **91**:509-514.

Lotufo, G.R., Landrum, P.F., Gedeon, M.L., Tigue, E.A., Herche, L.R. 2000. Comparative toxicity and toxikinetics of DDT and its major metabolites in freshwater amphipods. *Environmental Toxicology and Chemistry* **19**: 368-379.

MacDonald, D.D., Carr, R.S., Calder, F.D., Long, E.R., Ingersoll, C.G. 1996. development and evaluation of sediment quality guidelines for Florida Coastal Waters. *Ecotoxicology* **5**: 253-278.

Maltby, L., Blake, N., Brock, T.C.M. and Van den Brink, P. 2005. Insecticide Species Sensitivity Distribution: Importance of test species selection and relevance to aquatic ecosystems. *Environmental Toxicology and Chemistry* **24**: 379-388.

Maund, S.J., Hamer, M.J., Lane, M.C.G., Farrelly, E., Rapley, J H., Goggin, U.M., Gentle, W.E. 2002. Partitioning, bioavailability and toxicity of the pyrethroid insecticide cypermethrin in sediments. *Environmental Toxicology and Chemistry* **21**: 9-15.

McCall J.N., Fleeger, J.W. 1995. Predation by juvenile fish on hyperbenthic meiofauna: a review with data on postal-larval *Leiostomus xanthurus*. *Vie et Milieu* **45**: 61-73.

Mezquita, F., Hernandez, J.R., Rueda, J. 1999. Ecology and distribution of ostracods in a polluted Mediterranean river. *Paleogeography, Palaeoclimatology, Palaeoecology* **148**: 87-103.

Miliou, H. 1993. Temperature, salinity and light induced variations on larval survival and sex ration of *Tisbe holothuriae* Humes (Copepoda: Harpacticida). *Journal of Experimental Marine Biology and Ecology* **173**: 95-109.

Moore, M.T., Schulz, R., Cooper, C.M., Smith Jr, S., Rodgers Jr, J.H. 2002. Mitigation of chlorpyrifos runoff using constructed wetlands. *Chemosphere* **46**: 827-835.

Naidoo, V., Buckley, C.A. 2003. Survey of pesticide wastes in South Africa and review of treatment options. *Water Research Commission Report No. 1128/1/03*, Pretoria, South Africa.

Nozais, C., Perissinotto, R., Tita, G. 2005. Seasonal dynamics of meiofauna in a South African temporarily open/closed estuary (Mdloti Estuary, Indian Ocean). *Estuarine, Coastal and Shelf Science* **62**: 325-338.

Pait, A.S., DeSouza, A.E., Farrow, R.G. 1992. Agricultural pesticide use in coastal areas: A national summary. NOAA Strategic Environmental Assessments Division, Rockville, MD, USA.

Perissinotto, R., Walker, D.R., Webb, P., Wooldridge, T.H., Bailey, R. 2000. Relationships between zoo- and phytoplankton in a warm temperate, semi-permanently closed estuary, South Africa. *Estuarine, Coastal and Shelf Science* **51**: 1-11.

Perissinotto, R., Nozais, C., Kibirige, L. 2002. Spatio-temporal variations of phytoplankton and microphytobenthic biomass in a South African temporarily-open estuary. *Estuarine, Coastal and Shelf Science* **55**: 47-58.

Pequeux, A. 1995. Osmotic regulation in crustaceans. *Journal of Crustacean Biology* **16**: 1-60.

Pillay, D., Perissinotto, R. 2008. Community structure of epibenthic meiofauna in the St. Lucia Estuarine Lake (South Africa) during a drought phase. *Estuarine, Coastal and Shelf Science* **51**: 1-11.

Posthuma, L., Suter, G.W.II, Traas, T.P. 2002. Species-Sensitivity Distributions in Ecotoxicology. Lewis, Boca Raton, FL, USA.

Power, E. A., Chapman, P.M. 1992. Assessing Sediment Quality. In Sediment Toxicity Assessment. G. A. Burton, Jr. (ed.). Lewis Publishers, Inc., Boca Raton, FL. pp 1-18.

Preston, B.L. 2002. Indirect effects in aquatic ecotoxicology: Implications for ecological risk assessment. *Environmental Management* **29**: 311-323.

Quintino, V., Elliot, M., Rodrigues, A.M. 2006. The derivation, performance and role of univaraite and multivariate indicators of benthic change: Case studies at differing spatial scales. *Journal of Experimental Marine Biology and Ecology* **330**: 368-382.

Raisuddin, S., Kwok, K.W.H., Leung, K.M.Y., Schlenk, D., Lee, J.-S. 2007. The copepod Tigriopus: A promising marine model organism for ecotoxicology and environmental genomics. *Aquatic Toxicology* **83**: 161-173.

Ranasinghe, R., Pattiaratchi, C. 1999. Circulation and mixing characteristics of a seasonally open tidal inlet: a field study. *Marine and Freshwater Research* **50**: 281-290

Reddering, J.S.V. 1988. Prediction of the effects of reduced river discharge on estuaries of the south-eastern Cape Province, South Africa. *South African Journal of Science* **84**: 726-730.

Reinert, K.H., Giddings, J.M., Judd, L. 2002. Effects Analysis of Time-Varying or Repeated Exposures in Aquatic Ecological Risk Assessment of Agrochemicals. *Environmental Toxicology and Chemistry* **21** (9): 1977-1992.

Reuber, B., Mackay, D., Paterson, S., Stokes, P. 1987 A discussion of the chemical equilibria and transport at the sediment-water interface. *Environmental Toxicology and Chemistry* **6**: 731-739.

Sabljic, A., Gusten, H., Verhaar, H., Hermens, J. 1995. QSAR modelling of soil sorption. Improvements and systematics of $logK_{OC}$ versus $logK_{OW}$ correlations. *Chemopshere* **31**: 4489-4514.

Sánchez-Bayo, F. 2006. Comparative acute toxicity of organic pollutants and reference values for crustaceans. I. Branchiopoda, Copepoda and Ostracoda. *Environmental Pollution* **139**(3): 385-420.

Scharler, U.M., Baird, D. 2003. The influence of catchment management on salinity, nutrient stochiometry and phytoplankton biomass of Eastern Cape estuaries, South Africa. *Estuarine Coastal and Shelf Science* **56**: 735-748.

Schulz, R., Liess, M. 1999. A Field Study of the Effects of Agriculturally Derived Insecticide Input on Stream Macroinvertebrate Dynamics. *Aquatic Toxicology* **46**: 155-176

Schulz, R. 2001. Rainfall-induced sediment and pesticide input from orchards into the Lourens River, Western Cape, South Africa: Importance of a single event. *Water Research* **35**: 1869-1876.

Schulz, R., Liess, M. 2001. Toxicity of aqueous-phase and suspended particle-associated fenvalerate: chronic effects following pulse-dosed exposure of *Limnephilus lunatus* (Trichoptera *Environmental Toxicology and Chemistry* **20**:185-190.

Schulz, R., Peall, S.K.C. 2001. Effectiveness of a constructed wetland for retention of nonpoint-source pesticide pollution in the Lourens River Catchment, South Africa. *Environmental Science and Technology* **35**: 422-426.

Schulz, R, Peall, S.K.C, Dabrowski, J.M., Reinecke, A.J. 2001. Current-use insecticides, phosphates and suspended solids in the Lourens River, Western Cape, during the first rainfall event of the wet season. *Water SA* **27**: 65-70.

Schulz, R. 2003. Using a freshwater amphipod in situ bioassay as a sensitive tool to detect pesticide effects in the field. *Environmental Toxicology and Chemistry* **22**: 1172-1176.

Scott, G.I., Fulton, M.H., Moore, D.W., Wirth, E.F., Chandler, G.T., Key, P.B., Daugomah, J.W., Strzier, E.D., Devane, J., Clark, J.R., Lewis, M.A., Finley, D.B., Ellenberg, W., Karnaky Jr., K.J. 1999. Assessment of risk reduction strategies for the management of agricultural nonpoint source pesticide runoff in estuarine ecosystems. *Toxicology and Industrial Health* **15**: 200-213.

Selck, H., Palmqvist, A., Forbes, V.E. 2003. Uptake, depuration and toxicity of dissolved and sediment-bound fluoranthene in the polychaete, Capitella sp. I. *Environmental Toxicology and Chemistry* **22**: 2354-2363.

Segstro, M. D., Muir, D. C. G. Servos, M. R., Webster, G. R. B. 1995. Longterm fate and bioavailability of sediment-associated polychlorinated dibenzo-p-dioxins in aquatic mesocosms. *Environmental Toxicology and Chemistry* **14**(10): 1799-1807.

Shannon, C.E., Wiener, N., 1948. A mathematical theory of communication. Bell System *Technology Journal* **27**: 379-423.

Sherman, K.M., Coull, B.C. 1980 The response of meiofauna to sediment disturbance. *Journal of Experimental Marine Biology and Ecology* **45**: 59-71.

Solomon, K.R, Giddings J.M., Maund, S.J. 2001. Probabilistic risk assessment of cotton pyrethroids: I. Distributional analysis of laboratory aquatic toxicity data. *Environmental Toxicology and Chemistry* **20**: 652-659.

Soetaert, K., Vinex, M., Wittoeck, J., Tulkens, M., van Gansbeke, D. 1994. Spatial patterns of the Westerschelde meiobenthos. *Estuarine, Coastal and Shelf Science* **39**: 367-388.

Spargo, P.E., 1991. False Bay, South Africa – an historic scientific overview. *Transactions of the Royal Society of South Africa* 47: 363-375.

Sprague JB. 1985. Factors that modify toxicity. In Rand GM, Petrocelli SR, eds,

Fundamentals of Aquatic Toxicology. Hemisphere, New York, NY, USA, p. 124-163.

Staton, J.L., Schizas, N.V., Klosterhaus, S.L., Griffitt, R.J., Chandler, G.T., Coull, B.C. 2002. Effect of salinity variation and pesticide exposure on an estuarine harpacticoid copepod, *Microarthridion littorale* (Foppe), in the southeastern US. *Journal of Experimental Marine Biology and Ecology* **278**: 101-110.

Steen, R.J.C.A, Leonards, P.E.G., Brinkman, U.A.Th , Barcelo, D., Troncynski, J., Albanis, T.A., Cofino, W.P. 1999. Ecological risk assessment of agrichemicals in European estuaries. *Environmental Toxicology and Chemistry* **18**: 1574-1581.

Steen, R.J.C.A., van der Vaart, J., Hiep, M., van Huttum, B., Cofino, W.P., Brinkman, U.A.Th. 2001. Gross fluxes and estuarine behavior of pesticides in the Scheldt estuary (1995-1997). *Environmental Pollution* **115** 65-79.

Sujatha, C.H., Nair, S.M., Chacko, J. 1999. Determination and distribution of endosulfan and malathion in an Indian estuary. *Water Research* **33**: 109-114.

Sustek, Z. 1984. The bioindicative and prognostic significance of sex-ratio in Carabidae (Insecta, Coleoptera). *Ekologia* **3**, 3–22.

Suter, G.W., Barnthouse, L.W., Bartell, S.M., Mill, T., Mackay, D., Patterson, S. 1993. Ecological Risk Assessment. Lewis Publishers, Boca Raton.

Stucchi-Zucchi, A., Salomão, L.C. 1998. The ionic basis of membrane potentials and adaptation to hypcsmotic stress in *Perna perna*, an osmoconforming mollusk. *Comparative Biochemistry and Physiology* **121A**: 143-148.

Taljaard, S., van Ballegooyen, R.C., Morant, P.D. 2000. False Bay water quality review. Volume 2: Specialist assessment and inventories of available literature and data. Report to the False Bay Water Quality Advisory Committee. *CSIR Report ENV-S-2000-086/2*. Stellenbosch.

Ter Braak, C. J. F., Šmilauer, P. 2002. CANOCO Reference Manual and CanoDraw for Windows User's Guide: Software for Canonical Community Ordination (version 4.5). (Ithaca, New York, USA: Microcomputer Power)

Thiere, G., Schulz, R. 2004. Runoff-related agricultural impact in relation to macroinvertebrate communities of the Lourens River, South Africa. *Water Research* **38**(13): 3092-3102.

Turner, A., Millward, G.E. 2002. Suspended particles: Their role in estuarine biogeochemical Cycles. *Estuarine, Coastal and Shelf Science* **55**: 857-883.

Turpie, J.K., Adams, J.B., Joubert, A., Harrison, T.D., Colloty, B.M., Maree, R.C., Whitfield, A.K., Wooldridge, T.H., Lamberth, S.J., Taljaard, S. & van Niekerk, L. 2002. Assessment of the conservation priority status of South African estuaries for use in Heydorn, A.E.F. 1986. An assessment of the state of the estuaries of the Cape and Natal in 1985/6. South African National Scientific Programmes Report No. 130. Pretoria: CSIR. Management and water allocation. *Water SA* **28**: 191–206.

US EPA. 2002. Quality criteria for water 2002. EPA-822-R-02-047. Office of Water Regulations and Standard, Washington DC.

Van Beelen, P., Verbruggen, E.M.J., Peijnenburg, W.J.G.M. 2003. The evaluation of the equilibrium partitioning method using sensitivity distributions of species in water and soil. *Chemosphere* **53**: 1153-1162.

Van den Brink, P.J., Donk, E.V., Gylstra, R., Crum, S.J.H., Brock, T.C.M. 1995. Effects of Chronic Low Concentrations of the Pesticides Chlorpyrifos and Atrazine in Indoor Freshwater Microcosms. *Chemosphere* **31** (5): 3181-3200.

Van den Brink, P.J., Ter Braak, C.J.F. 1999 Principal Response Curves: Analysis of Time-dependent Multivariate Responses of a Biological Community to Stress. *Environmental Toxicology and Chemistry* **18**: 138-148.

Van den Brink, P.J., Van den Brink, N.W., Ter Braak, C.J.F. 2003. Multivariate analysis of ecotoxicological data using ordination: Demonstrations of utility on the basis of various examples. *Australasian Journal of Ecotoxicology* **9**: 141-156.

Van den Brink, P.J., Den Besten, P.J., Bij de Vaate, A., ter Braak, C.J.F. 2009. The use of the Principal Response Curves techniques for the analysis of multivariate time series from biomonitoring studies. *Environmental Monitoring and Assessment.* (in press)

Valette-Silver, J.N., Brown L., Pavich, M., Klein, J., Middleton, R. 1986. Detection of erosion event using 10Be profiles: example of the impact of agriculture on soil erosion in Chesapeake Bay area (U.S.A). *Earth and Planetary Science Letters* **80**: 82-90.

Villa, S., Finizio, A., Vighi, M. 2003a. Pesticide risk assessment in a Lagoon ecosystem. Part 1: Exposure Assessment. *Environmental Toxicology and Chemistry* **22**: 928-935.

Villa, S., Vighi, M., Casini, S., Focardi, S. 2003b. Pesticide Risk Assessment in a Lagoon Ecosystem. Part II: Effect Assessment and Risk Characterization. *Environmental Toxicology and Chemistry* **22**: 936-942.

Voice, T. C. and W. J. Weber, Jr. 1983. Sorption of hydrophobic compounds by sediments, soils, and suspended solids-I: Theory and background. *Water Research* **17**(10): 1433-1441.

Wan, M.T., Kuo, J.K., Buday, C., Schroeder, G., Van Aggelen, G., Pasternak, J. 2005. Toxicity of α-, β-, (α+β)-endosulfan and their formulated and degradation products to *Daphnia magna*, *Hyalella azteca*, *Oncorhynchus mykiss*, *Oncorhynchus kisutch*, and biological implications in streams. *Environmental Toxicology and Chemistry* **24**: 1146-1154.

Warwick, R.M., Platt, H.M., Clarke, K.R., Agard, J., Gobin, J. 1990. Analysis of macrobenthic and meiobenthic community structure in relation to pollution and disturbance in Hamilton Harbour, Bermuda. *Journal of Experimental Marine Biology and Ecology* **138**: 119-142.

Warwick, R.M., Clarke, K.R. 1993. Increased variability as a symptom of stress in marine communities. *Journal of Experimental Marine Biology and Ecology* **172**: 215-226.

Weinstein, J.E. 2003. Influence of salinity on the bioaccumulation and photoinduced toxicity of fluoranthene to an estuarine shrimp and oligochaete. *Environmental Toxicology and Chemistry* **22**: 2932-2939.

Weston, D.P., Penry, D.L., Gulman, L.K. 2000. The role of ingestion as a route of contaminant bioaccumulation in a deposit-feeding polychaete. *Archive of Environmental Contamination and Toxicology* **38**: 446-454

Wheeler, J.R., Leung, K.M.Y., Morritt, D., Sorokin, N., Rodgers, H., Toy, R., Holt, M., Whitehouse, P., Crane, M. 2002. Freshwater to saltwater toxicity extrapolation using species sensitivity distributions. *Environmental Toxicology and Chemistry* **21**: 2459-2467.

Whitfield, A.K. 1992. A characterization of Southern African estuarine systems. *South African Journal of Aquatic Sciences* **18**: 89-103.

Whitfield, A.K., Bate, G. 2007. The freshwater requirements of intermittently open Cape estuaries. *WRC Report No 1581/1/07*

Willmer, P., Stone, G., Johnston, L. 2000. Environmental Physiology of Animals. Blackwell Science, Oxford., p. 644.

References

Wooldridge, T.H. & Callahan, R. 2000. The effects of a single freshwater release into the Kromme estuary. 3: Estuarine zooplankton response. *Water SA* **26** (3): 311-318.

Zhang, Z.L., Hong, H.S., Zhou, J.L., Huang, J., Yu, G. 2003. Fate and assessment of persistent organic pollutants in water and sediment from Minjiang River estuary, Southeastern China. *Chemosphere* **52**: 1423-1430.

Zulin, Z., Huasheng, H., Xinhong, W., Jianqing, L., Weiqi, C., Li,. X. 2002. Deteremination and load of organophosphorus and organochlorine pesticides at water from Jiulong River Estuary, China. *Marine Pollution Bulletin* **45**: 397-40

Die VDM Verlagsservicegesellschaft sucht für wissenschaftliche Verlage abgeschlossene und herausragende

Dissertationen, Habilitationen, Diplomarbeiten, Master Theses, Magisterarbeiten usw.

für die kostenlose Publikation als Fachbuch.

Sie verfügen über eine Arbeit, die hohen inhaltlichen und formalen Ansprüchen genügt, und haben Interesse an einer honorarvergüteten Publikation?

Dann senden Sie bitte erste Informationen über sich und Ihre Arbeit per Email an *info@vdm-vsg.de*.

Sie erhalten kurzfristig unser Feedback!

VDM Verlagsservicegesellschaft mbH
Dudweiler Landstr. 99
D - 66123 Saarbrücken
www.vdm-vsg.de

Telefon +49 681 3720 174
Fax   +49 681 3720 1749

Die VDM Verlagsservicegesellschaft mbH vertritt

Printed by Books on Demand GmbH, Norderstedt / Germany